Design and Use of Assisti

Meeko Mitsuko K. Oishi • Ian M. Mitchell
H. F. Machiel Van der Loos
Editors

Design and Use of Assistive Technology

Social, Technical, Ethical, and Economic Challenges

Foreword by Maja J. Matarić

 Springer

Editors

Meeko Mitsuko K. Oishi
University of British Columbia
Electrical and Computer
Engineering
2332 Main Mall
Vancouver, BC V6T 1Z4
Canada
moishi@ece.ubc.ca

H. F. Machiel Van der Loos
University of British Columbia
Department of Mechanical Engineering
6250 Applied Science Lane
Vancouver, BC V6T 1Z4
Canada
vdl@mech.ubc.ca

Ian M. Mitchell
University of British Columbia
Department of Computer Science
201-2366 Main Mall
Vancouver, BC V6T 1Z4
Canada
mitchell@cs.ubc.ca

ISBN 978-1-4899-8980-2   ISBN 978-1-4419-7031-2 (eBook)
DOI 10.1007/978-1-4419-7031-2
Springer New York Dordrecht Heidelberg London

Foreword

The not too distant future will feature assistive technology as an integral part of people's lives. This future is ensured by the disquieting speed at which the need for that technology is growing. With the aging population trends in developed countries and the growing rates of developmental and other disorders and conditions in children, large sectors of the population are in need of one-on-one, dedicated, and individualized care. At the same time, our species' ever-increasing lifespan means such care is needed for increasingly longer periods, beyond convalescence and into long-term rehabilitation and life-long support. Given population demographics, there simply will not be enough human labor available to provide this needed care. The resulting "care gap" presents a niche for human-centered technology. What does it take to create such technology?

The challenges of safe, ethical, culturally-appropriate, engaging, accessible, and affordable assistive technology are many, and constitute the motivating forces of the growing interdisciplinary research trend reflected in this book. If we are to make progress toward addressing these challenges, we must create a culture that sustains an active interaction between the technology developers and the intended user communities, by bringing them much closer together than they have been or have needed to be so far. The two communities must inform each other; the collaboration must be bidirectional. In an effective collaboration, users direct research to make it relevant to real-world needs, and researchers disseminate technologies and training through service models. To succeed, we must also address the complex ethical issues of technology acceptance, dependence, access, and safety throughout the process, not after the technology is developed. Finally, the foundation for making all this possible must rest on a funding structure that provides support for the full cycle of collaborative, interdisciplinary, and community-centered development, testing, evaluation, and iterative improvement.

The future is in sight but we are not yet there. This book and related work aid in outlining the steps that will bring us closer to what will be a true

paradigm shift in technology-aided medicine and health. In the words of a
participant in one of our assistive technology studies, "The sooner we can do
this, the better!"

University of Southern California   *Maja J. Matarić*
May 2010

Preface

Assistive technologies have the potential to significantly improve the lives of people with disabilities, by enabling independence and facilitating social connections. However, these same technologies can be a barrier to independence and social connectedness if they are poorly designed, do not effectively incorporate user requirements, or are inappropriate for the task at hand. Effective assistive technologies depend not only on "good" engineering design (sometimes a challenge in and of itself), but also on the extent to which the technology has been integrated with clinical needs, user requirements, ethical concerns, and the social context of the technology's use. In fact, poor, ineffective, or inappropriate design is a key cause of device abandonment. The gaps between engineering design, clinical evaluation, and actual use represent an inherently multidisciplinary challenge that must be forcefully and creatively addressed if assistive technologies are to better succeed in enhancing people's lives. In a workshop held July 22–24, 2009 at the University of British Columbia, "Removing barriers and enabling individuals: Ethics, design, and use of assistive technology" (http://www.removingbarriers.pwias.ubc.ca), clinicians and researchers from computer science, engineering, ethics, medicine, and rehabilitation sciences gathered specifically to address this increasingly expensive issue.

This book is a result of presentations and discussions that took place over the course of the 3-day joint Peter Wall Institute for Advanced Studies (PWIAS)/Institute for Computing, Information, and Cognitive Systems (ICICS) Exploratory Workshop. The workshop was unique in its interdisciplinary focus and opportunities for multidisciplinary, small-group discussion. The workshop focused on four different *topics*: evaluation, sensing, networking, and mobility, and four different cross-cutting *themes*: novelty, customization, privacy, and user perspective. Workshop participants were encouraged to go beyond mere anecdote to identification of current problems and potential improvements in the entire cycle of design, evaluation, knowledge transfer, and actual use of assistive technologies. Three key recommendations were identified: (1) The user's experience must be fully – not just partially, anecdotally or vicariously – integrated into both engineering design and clinical evaluation. (2) Academic outreach via service models should be widely adopted to create customized assistive technology solutions for clients as the experience simultaneously educates the embedded researcher and students in real-world situations. (3) Knowledge transfer should be enabled through the creation and enforcement of regulations and standards to increase quality and reduce cost, the implementation of mechanisms to pool risk and limit liability, and the financial support to small businesses to capitalize on the inherent advantages of agile, niche companies.

The purpose of this book is to assess some of the major hurdles in bringing assistive technologies out of the lab and into everyday use and to provide guidelines and recommendations to improve their design and use. Some of the most difficult problems in creating effective assistive technology are (a) the inherent heterogeneity of the user population, (b) privacy concerns in data gathering and analysis, (c) knowledge transfer of novel technologies, and (d) incorporation of user perspective into the design process. It is our belief that true solutions to these issues can only arise through a multidisciplinary approach.

We have gathered in this book a set of papers that demonstrate how process improvements in assistive technology deployment have the potential to empower businesses, researchers, and nonprofit organizations to create and bring to market new devices, such that they incorporate ethical, social, and clinical concerns by design. Contributors to the book are leading researchers in their fields, and their contributions are inherently broad in scope and accessible to researchers from a wide range of disciplines. We provide a critical assessment of hurdles in assistive technology that are relevant for researchers in engineering, computer science, rehabilitation sciences, and ethics. The book is organized according to the main outcomes of the workshop: regarding the user's experience, research and academic outreach, and development and commercialization.

We begin with a discussion of the issues that inherently frame how assistive technology is conceived of, designed, used, and perceived. In Chap. 1, Silvers identifies key ways in which ethics of assistive technology differ from seemingly similar issues in engineering ethics and medical ethics. Miller Polgar provides an alternative framework in Chap. 2 in which to consider the role and assumed neutrality of assistive technology, both as a barrier and as an enabler. Even the words commonly used to describe "assistive" technology have implicit assumptions about perception of and identification as an individual with a disability. Ladner makes a case for "accessible" technology in Chap. 3. In Chap. 4, Cook and Adams focus on how technology can enable play for children with disabilities. In Chap. 5, Cook, Miller Polgar, and Livingston discuss need-based and task-based assistive technology design and evaluation as a means to prevent device abandonment. They employ the human activity assistive technology (HAAT) model to evaluate both successful and failed technologies.

The second part of the book focuses on models of the research pipeline and the role of academic outreach in improving how assistive technology is designed, evaluated, and used. Simpson discusses in Chap. 6 common barriers in the typical research pipeline. Livingston discusses in Chap. 7 how community service in academia can not only enable improved technology design and use, but is also a means to create a new workforce of engineers and technologists cognizant of accessibility issues, irrespective of whether the technology they design is truly "assistive." Matsuoka and Lewis provide a case study in Chap. 8 of the creation of a non-profit organization, spun off from work done originally in academia, to create highly-customized assistive technologies. Lastly, Danielson, Longstaff, Ahmad, Van der Loos, Mitchell, and Oishi discuss in Chap. 9 the results of a recent survey in the ethics of assistive technology that highlights some of the unique challenges in the research, development, evaluation, implementation, and use of assistive technologies.

The last section of the book discusses some of the most difficult aspects of improving assistive technologies – the broader legal and economic context that influences the development and commercialization of assistive technologies. In Chap. 10, Birch evaluates the current regulations and standards (as well as those in process but not yet implemented) and argues that enforcement of regulations and standards is required to provide truly universal access. Borisoff draws upon his personal experience as an entrepreneur, to discuss in Chap. 11 some of the unique challenges and opportunities in assistive technologies due to the small market for AT products.

We are pleased to acknowledge financial support from the Peter Wall Institute for Advanced Studies at the University of British Columbia (UBC), the UBC Institute for Computing, Information and Cognitive Systems, and the British Columbia Disabilities Health Research Network. The workshop would not have been possible without guidance and support from our Advisory Committee, a multidisciplinary team of leading researchers at UBC focused on the ethics, design, and use of assistive technologies. In addition, we are grateful to Dr. Dianne Newell, the Director of PWIAS, for her continued interest and support, and to her staff for their assistance.

University of British Columbia at Vancouver   *Meeko M.K. Oishi*
May 2010   *Ian M. Mitchell*
                                                       H.F. Machiel Van der Loos

Acknowledgments

UBC Advisory Committee

Elizabeth Croft, Ph.D.
Mechanical Engineering, Faculty of Applied Science

Peter Danielson, Ph.D.
Applied Ethics, W. Maurice Young Centre for Applied Ethics

Judy Illes, Ph.D.
Neuroscience, Faculty of Medicine

Alan Mackworth, Ph.D.
Computer Science, Faculty of Science

William C. Miller, O.T., Ph.D
Occupational Science and Occupational Therapy, Faculty of Medicine

Bonita Sawatzky, Ph.D.
Orthopedics, Faculty of Medicine

PWIAS-ICICS Exploratory Workshop Participants

Mark Ansermino, M.D.
Gary Birch, O.C., Ph.D., P. Eng.
Jean-Sébastien Blouin, Ph.D.
Jaimie F. Borisoff, Ph.D.
Anna Cavender
Albert M. Cook, Ph.D., P. Eng.
Susan Crawford
Peter Danielson, Ph.D.
Kirsty Dickinson
Antony Hodgson, Ph.D., P. Eng.
Thomas Huryn
Judy Illes, Ph.D.
Chandrika Jayant
Philippe Kruchten, Ph.D., P. Eng.
Richard Ladner, Ph.D.
Linda Lanyon, Ph.D.
Nigel J. Livingston, Ph.D.
Holly Longstaff
Alan M. Mackworth, Ph.D.
Yoky Matsuoka, Ph.D.
Johanne Mattie, M.A.Sc.
William C. Miller, Ph.D., O.T.
Ian M. Mitchell, Ph.D.
Meeko M.K. Oishi, Ph.D.
Wayne Pogue
Jan Miller Polgar, Ph.D., O.T.
Bonita Sawatzky, Ph.D.
Richard Simpson, Ph.D., O.T.
Chris Speropoulos
H.F. Machiel Van der Loos, Ph.D.
Rongrong Wang

Contents

Part III Development and Commercialization

List of Contributors

Kim Adams, Ph.D., P. Eng.
Faculty of Rehabilitation Medicine, University of Alberta, Edmonton, AB, Canada, and Glenrose Rehabilitation Hospital, 10230-111 Avenue, Edmonton, AB, T5G OB7, Canada, e-mail: kim.adams@ualberta.ca

Rana Ahmad
Philosphy, University of British Columbia, Vancouver, BC, V6T 1Z4, Canada, e-mail: rahmad@interchange.ubc.ca

Gary Birch, O.C., Ph.D., P. Eng.
Neil Squire Society, 220-2250 Boundary Road, Burnaby, BC, V5M 3Z3, Canada, e-mail: garyb@neilsquire.ca

Jaimie F. Borisoff, Ph.D.
Instinct Mobility Inc., and Brain Interface Lab, Neil Squire Society, International Collaboration on Repair Discoveries, 818 West 10th Avenue, Vancouver, BC, V5Z 1M9, Canada, e-mail: jaimieb@gmail.com

Albert M. Cook, Ph.D.
Speech Pathology & Audiology, Faculty of Rehabilitation Medicine, University of Alberta, Corbett Hall, Edmonton, AB, T6G 2G4, Canada, e-mail: al.cook@ualberta.ca

Peter A. Danielson, Ph.D.
W. Maurice Young Centre for Applied Ethics, University of British Columbia, 227-6356 Agricultural Road, Vancouver, BC, V6T 1Z2, Canada, e-mail: pad@ethics.ubc.ca

Richard E. Ladner, Ph.D.
Computer Science & Engineering, University of Washington, Box 352350, Seattle, WA 98104, USA, e-mail: ladner@cs.washington.edu

Brian E. Lewis, J.D.
Rosen Lewis PLLC, 615 Second Avenue, Suite 760, Seattle, WA, 98104, USA, e-mail: blewis@rosenlewis.com

Nigel J. Livingston, Ph.D.
CanAssist, University of Victoria, PO Box 1700, STN CSC, Victoria,
BC, V8W 2Y2, Canada, e-mail: njl@uvic.ca

Holly Longstaff
Interdisciplinary Studies, University of British Columbia, 164-1855 West
Mall, Vancouver, BC, V6T 1Z2, Canada, e-mail: longstaf@interchange.
ubc.ca

Yoky Matsuoka, Ph.D.
Computer Science & Engineering, University of Washington, Box 352350,
Seattle, WA 98195, and YokyWorks Foundation, 6513 132nd Ave. NE #385,
Kirkland, WA 98033, e-mail: yoky@cs.washington.edu

Ian M. Mitchell, Ph.D.
Dept. Computer Science, University of British Columbia, 201-2366 Main
Mall, Vancouver, BC, V6T 1Z4, Canada, e-mail: mitchell@cs.ubc.ca

Meeko M. K. Oishi, Ph.D.
Electrical and Computer Engineering, University of British Columbia,
2332 Main Mall, Vancouver, BC, V6T 1Z4, Canada,
e-mail: moishi@ece.ubc.ca

Jan Miller Polgar, Ph.D., O.T. Reg. (Ont.) FCAOT,
School of Occupational Therapy, Elborn College, 1201 Western Road,
The University of Western Ontario, London, ON, N6G 1H1, Canada,
e-mail: jpolgar@uwo.ca

Anita Silvers, Ph.D.
Philosophy, College of Humanities, San Francisco State University, 1600
Holloway Avenue, San Francisco, CA, 94132, e-mail: asilvers@sfsu.edu

Richard Simpson, Ph.D., O.T.
School of Health & Rehabilitation Science, University of Pittsburgh, 4020
Forbes Tower, Pittsburgh, PA, 15260, e-mail: ris20@pitt.edu

H.F. Machiel Van der Loos, Ph.D.
Dept. Mechanical Engineering, University of British Columbia, 6250 Applied
Science Lane, Vancouver, BC, V6T 1Z4, Canada, e-mail: vdl@mech.ubc.ca

Acronyms and Abbreviations

AAC	Augmentative and Alternative Communication
ACM	Association for Computing Machinery
ADA	Americans with Disabilities Act
ANSI	American National Standards Institute
ASL	American Sign Language
ASSETS	ACM SIGACCESS Conference on Computers and Accessibility
AT	Assistive Technology
CAE	Centre for Applied Ethics
CAOT	Canadian Association of Occupational Therapists
CAPTCHA	Completely Automated Public Turing test to tell Computers and Humans Apart
CHI	International Conference on Human Factors in Computing Systems
CRTC	Canadian Radio–television and Telecommunications Commission
DHRN	British Columbia Disabilities Health Research Network
DSS	DriveSafe System
EADL	Electronic Aid to Daily Living
EATS	Efficiency of Assistive Technology and Services
EMG	Electromyography
E&J	Everest and Jennings, Inc.
FCAOT	Fellow of the Canadian Association of Occupational Therapists
FCC	US Federal Communications Commission
FDA	US Food and Drug Administration
G3ict	Global Initiative for Inclusive Information and Communication Technologies
GPS	Global Positioning System
HAAT	Human Activity Assistive Technology
HAC	Hearing Aid Compatibility

HCI	Human–Computer Interaction
ICICS	UBC Institute for Computing, Information, & Cognitive Systems
ICRA	IEEE International Conference on Robotics and Automation
ICT	Information and Communication Technologies
IDEA	US Individuals with Disabilities Education Act
IEEE	Institute for Electrical and Electronics Engineers
IEP	Individualized Education Plan
IRD	Interactive Robotic Device
IROMEC	Interactive Robotic social Mediators as Companions
ISAAC	International Society for Augmentative and Alternative Communication
MBA	Master of Business Administration
MPT	Matching Person and Technology Assessment
NASA	US National Association for Space and Aeronautics
O.C.	Order of Canada
OCR	Optical Character Recognition
OT	Occupational Therapist
O&M	Orientation and mobility
PIADS	Psychosocial Impact of Assistive Devices Scale
PWIAS	UBC Peter Wall Institute for Advanced Studies
P.Eng.	Professional Engineer
QUEST	Québec User Evaluation of Satisfaction with Assistive Technology
RESNA	Rehabilitation Engineering and Assistive Technology Society of North America
SIGACCESS	Special Interest Group on Accessible Computing
SGD	Speech Generating Device
STEM	Science, Technology, Engineering, and Mathematics
TADNSW	Technical Aid to the Disabled of New South Wales
UBC	University of British Columbia
UIST	User Interface Software and Technology
UN	United Nations
UVATT	University of Victoria Assistive Technology Team
VA	US Dept. of Veterans Affairs
WCAG	Web Content Accessibility Guidelines
WSP	Wireless Service Provider
W3C	World Wide Web Consortium

Part I
The User's Experience

Chapter 1
Better Than New! Ethics for Assistive Technologists

Anita Silvers

Abstract What are the fundamental values that should guide the practice of assistive technologists? This essay examines two sources that appear to inform current understandings of the ethics of assistive technology: medical ethics and engineering ethics. From medical ethics comes the notion that assistive technology should aim to restore its users to normal functioning, making them like new. Engineering ethics, on the other hand, recommends enhancing users' functionality, even if functioning is not achieved in a species typical way. From this engineering perspective, it is permissible and even desirable for assistive technology to make its users function even better than new. Thus enhancing functionality is a central value in assistive technology. Professionals in the field have the ability, and the responsibility as well, to address and counter societal suspicion of artificially enhanced functioning achieved through technology. Consequently, assistive technology professionals should fight against discrimination that excludes people with disabilities, whose functioning depends on prostheses and other products of technology, from the mainstream of social life.

1.1 Introduction

As is common at the commencement of a new profession, the ethics of assistive technology is a somewhat discordant conjunction drawn from other practices – in this case, from engineering ethics and also from medical ethics. The emphasis in the current approach is on doing no harm, prompted by familiar fears about the propensity of new technologies to trigger unfortunate results [5]. This kind of "hand-me-down" approach to shaping the aspirations,

Anita Silvers
San Francisco State University, San Francisco, CA, USA, `asilvers@sfsu.edu`

M.M.K. Oishi et al. (eds.), *Design and Use of Assistive Technology: Social,*
Technical, Ethical, and Economic Challenges, DOI 10.1007/978-1-4419-7031-2_1,
© Springer Science+Business Media, LLC 2010

as well the conduct, of assistive technologists may have the effect of delaying the maturation of the field, not to mention disrespecting the autonomy of and deferring justice for disabled people.

The unreflective importation of the values of rehabilitation, which aims to repair or restore people, making them as normal as possible and ideally "like new," is problematic. This is an understandable but unworthy goal for assistive technologists, whose ethics should be grounded in more innovative ideas about how people function and society may progress. In what follows I explain the current sources of assistive technology ethics. Then I argue that functioning "better than new" is as good a goal for assistive technology as functioning "like new." From this it follows that the goal of functioning "other than like new" is appropriate and acceptable in developing assistive technology. Assistive technologists should give the value of "functioning as well as possible" precedence over the value of "functioning normally" and should not be concerned if the outcome is to give assistive technology users functional advantage in some respects.

1.2 Fear of Technology and Disability Discrimination

The repugnant intervention of humans who impose technology to corrupt nature's design, or attempt to override naturally imposed limitations by trying to control and improve the effects of natural processes, has been a powerfully influential cultural theme at least since the Romantic movement of the nineteenth century. The stories of the creation by Frankenstein of an artificial man, and of Hyde by Jekyl, are well-known progenitors of an enormous number of works in the romantic genre. Romanticism privileges nature and its products (regardless of inadequate, unfair or destructive natural processes) over what humans craft.

While this way of looking at the world dominates our understanding of the human condition, the idea of the natural usually has been elided with the standard of the normal. And the identification of normality in humans usually is reduced to species typicality, so what is typical of humans is equated with what is normal for us. Thinking in the "romanticist" style thus has caused enormous harm to individuals who function atypically or anomalously by making unusual people seem "abnormal" and consequently dangerously deviant, rather than merely different. Individuals evaluated as abnormal are too easily supposed to disrupt, threaten or burden the personal or social lives of normal ones.

Individuals with disabilities often are ostracized because they are imagined to be less functional and therefore weaker than other people. But a contrary rationale becomes operative when assistive technology enters the picture. Societal suspicion of artificially enhanced functioning achieved through technology sometimes prompts the exclusion of disabled individuals who rely on prosthetics to participate in the mainstream of social life. Examples of such prejudice are numerous, from denying blind law students use of common screen reading technology to take the bar exam[1] to excluding a competitor with lower limb prostheses from the Olympic games. A third example, more widespread, is anxiety about danger to pedestrians, which is directed at wheelchair users.

In general, it is not unusual for prosthetics that enable people with disabilities to function independently and competitively to be condemned prospectively for their supposed potential for social destabilization. In the first case mentioned above, the California Bar Association was comfortable with the applicant using a screen reader but the company marketing the bar exam feared that the screen reading software would steal the exam questions (which are reused for several exams).[2] The National Conference of Bar Examiners, the vendor for part of the exam, objected to the use of standard screen reading technology, claiming (among other reasons) that the bar exam items could be stolen, but this expression of fear did not prevail. In the second case, the International Association of Athletics Federations objected on the ground that the lower limb prosthetics user did not exert as much below-the-knee effort as runners on flesh feet,[3] but was overruled by the Court of Arbitration for Sport, which found no evidence that the prosthetics user expended less energy overall because of his artificial running feet. (I will return to the third case – conflict over safety between wheelchair using person and walking people – in the Afterword.)

A commonly expressed fear is that using technology might catapult a formerly disadvantaged person to advantaged status, altering who is positioned where on the social playing field. Both of the cases referenced above have this dimension. In the first case, the vendor of the national segment of the exam also objected to the plaintiff's using a screen reading program, claiming that all blind bar exam takers might request this accommodation.[4] In the second case, track officials speculated that permitting an individual with a disability to compete on manufactured racing feet would open the way to track competitors being allowed many other kinds of assistive devices, including skates.

---

[1] See [16] for the latest US court decision permitting use of screen reading technology.

[2] Apparently, the company did not think the human reader it proposed to hire long distance would be as larcenously inclined.

[3] In early commentary, international track officials invoked a range of speculative claims about the destabilizing effects of allowing a double amputee using manufactured feet to compete. These included fears that he would harm other runners by falling over on them and that other athletes would have their legs amputated so as to be fitted with artificial racing feet. I have addressed these exaggerations and the prejudices leading to them in [13].

[4] Why the specter of a blind test taker independently using screen reading programs, rather than depending on human readers of uncertain quality plus Braille, was presented by the vendor as a harm remains obscure, except for the suggestion that allowing screen reading programs would give more blind individuals access to bar exams and thus to legal careers.

In general, allowing disabled individuals the same opportunity to participate as others when their functioning involves technology reduces the underclass of deficient individuals, leaving normal people a smaller number to whom they can feel superior.

There, thus, is a widespread social habit of rejecting individuals with disabilities based on their intimate ties with the products of technology. Assistive technology professionals ought to be aware of this familiar reaction, and most are. They also ought to understand this prejudice for what it is and be committed to facing and dealing with it. Unfortunately, the sources available from which to draw guidance in professional ethics for assistive technology do not prepare them to do so.

1.3 Sources of Assistive Technology Ethics

1.3.1 Engineering Ethics

Codes of engineering ethics[5] began to be developed early in the twentieth century, coincident with engineers organizing as a profession sufficiently responsible to regulate itself. Unlike medical ethics, where individual patients are regarded as the primary beneficiaries, engineers are held first of all to promote the public or general good. They must protect public health, safety and welfare, sometimes against the interests of employers, to whom they have an obligation of loyalty. Harm caused by structural failure sometimes, although not always, arises from design or construction failure to which ethical failures, such as conflict of interest or bribery or deflection of responsibility or absence of respect for other humans, have contributed.

As the twentieth century drew to a close, engineers found themselves working with structures of kinds unimaginable to their predecessors at the beginning of the century – not only in electrical engineering but most prominently in bioengineering. Because the structures bioengineers create enable individual humans to achieve biological functioning, albeit through mechanical or electrical means, for bioengineers, including designers of assistive technology, the ethics of bioengineering resides at the intersect of medical and engineering ethics.

For example, contrary to some prevailing practice, the medical devices they design should be tested on human subjects, with the risk-benefit standards and safeguards common to human subject research, such that subjects:

---

[5] For examples of engineering codes of ethics and discussions of ethical dilemmas specific to engineering, see the excellent website of the National Academy of Engineering, Online Ethics Center for Engineering and Research [4]. See also the Code of Ethics of the Rehabilitation Engineering and Assistive Technology Society of North America [6].

1. Are honestly informed about risks
2. Are potentially benefited from the research, either personally or by bringing about a more general good that gratifies them
3. Are competent to weigh personal risks and benefits
4. Are specially protected from exploitation if they belong to historically vulnerable groups, and people with disabilities belong to one or more such vulnerable groups.

These safeguards should be extrapolated to use contexts, as users may be supposed to be exposed to some of the same risks (in order to achieve the same personal benefits) as the test subjects. Currently, some assistive technology items are not submitted to human subjects testing, which is standard for prescription drugs.[6] Their evaluation is more like that of bridges and laptops, a combination of construction and product standards with marketing research. Further, prospective users often are not accorded the opportunity to be informed of the potential risks and benefits of using each assistive or prosthetic device, and to consent, other than to be warned about (mis)uses of devices in attempts to defend against product liability.

1.3.2 Medical Ethics

Many assistive and prosthetic devices are acquired through medical prescription. As engineers move into designing devices that are distributed in this way,

---

[6] See [17] for an easily understood description of the distinction between FDA requirements for Class 3 devices (usually invasive, such as implantable defibrillators) and Class 2 devices such as wheelchairs and communication devices. A review of FDA procedures by the Institute of Medicine is due in Spring 2011.

For an account of lack of testing in, as an example, wheelchair design, and production see [2]. Cooper writes here:

"Unfortunately, neither the VA nor FDA, or any other US-based agency made the standards a requirement. This has resulted in a virtual flea market for testing, as manufacturers and distributors pick and choose which tests to apply, create their own tests, or even choose to ignore the issue altogether. In the end, wheelchair users, their families, and their caregivers pay the price."

See also Alberta Floyd, Individually and as Administrator of the Estate of Jacqueline Ann Adams, Deceased, et al., Plaintiffs, v. Pride Mobility Products Corp., Defendant. No. 1:05-CV-00389, United States District Court For The Southern District Of Ohio, Western Division, 2007 US Dist. Lexis 91287 as an example of court assessments of the nondispositive role of compliance with RESNA and ANSI standards in regard to wheelchair design safety issues.

See also [15], a policy document that urges expansion of research on and manufacture of assistive technology products but does not address testing.

Of course, the research of university-based assistive technology researchers and developers is monitored by institutional research and human subjects committees, but universities are not central players in marketing assistive technology products.

the ethical status of end-users of their products becomes more complex.[7] This is due not to technology users being exceptional or idiosyncratic but because of conflict in the current characterization of service receivers in the health care system.

On the one hand, the traditional portrayal of patients in medicine has been as unknowledgeable dependents who ought to be compliant with the expert decisions of physicians for their own good. On the other hand, courts adjudicating cases in which patients have not been benefitted but instead harmed, and legislators enacting statutes to prevent more such cases, have conceived of patients as autonomous agents who typically are competent to consent to risk for the sake of benefit (or to refrain from consenting) if only they are given appropriate information. The medical professionals' obligation to obtain patients' informed consent is grounded in this recently expanded notion of the doctor–patient relationship as like a contract between two individuals with equivalently respectable agency.

The development of codes or schemes of, or philosophical approaches to, professional ethics in medicine and related fields – endeavors that conduct medical research on human subjects or provide therapeutic or other health care services to them – became of increasing concern during the last half of the twentieth century. This effort responded to a deepening erosion of trust in medical professionals, fueled at least in part by the 1946–1947 trial of the Nazi doctors who sterilized, euthanized, and experimented on patients. Revelations about scandalous conduct toward patients entrusted to physicians' care exacerbated the evaporation of confidence in physicians giving the individual patients' good priority over other values that may drive them. Investigations at Tuskegee, Willowbrook, and other venues revealed that the health of patients whose welfare was placed in the hands of physicians had instead been sacrificed by those same physicians, who invoked the value of advancing scientific knowledge to defend their actions.

Restraints on professionals in whom the well-being of vulnerable patients was vested needed to be inherent in the healthcare professions themselves, or else the state would intervene in the forms of regulative legislation or restrictive case law. Codes of professional ethics partly staved off regulation through law. Nevertheless, this was a time of unprecedented statutory and judicial action aimed at protecting patients by emphasizing their right to determine whether to accept medical interventions.

Being subject to regulatory law, and to civil rights protections for patients, altered the conceptual framework of medicine. In the courts at least, if not as clearly and consistently in the clinic, patients no longer were conceptualized as passive and clueless beings dependent on the benevolence of healthcare professionals but were thought of, instead, as (usually) competent choosers with the right to control what is done to their bodies and their minds.

---

[7] The descriptions in this section are familiar ones in the literature of medical ethics. For expanded discussion of them, see such standard textbooks as [1, 14, 7].

In recent years, therefore, the status of the patient in relation to the physician or other health care professional has come to resemble that of the client in relation to the engineer: a competent chooser.

1.3.3 At the Intersect of Medical and Engineering Ethics

A first step for developing assistive technology ethics is to consider what engineering and medical ethics might have in common. Both center on directives to avert the kinds of harm that professionals in the field have special opportunity, occasion or opening to do. But engineering and medical ethics presume very different notions of who or what must be protected from harm. Engineering ethics aims to defend the general public welfare against harm caused by inadequate engineers. But medical ethics is focused on protecting individual patients from being harmed by feckless physicians.

Engineering ethics does not evidence similar worry about obligations to dependent individuals. Rather, individuals to whom ethical obligation is owed are cast mainly in the role of independent employers or clients. Clients are presumed autonomous in the sense of their being sufficiently powerful to control or manage their own fates. They are owed the loyalty of the marketplace – that is, compliance with their wishes by those whom they compensate for doing so. Engineering ethics does conceive of the employers' good as potentially in conflict with the public good, and ethical sensitivity and response to such conflict is a central issue for engineering ethics.

Medical ethics does not usually conceive of, and thus is not centrally concerned about, conflicts between the patients' good and the public good. The main exceptions are (1) where medical resources are so scarce as to require rationing and some kinds of patients are thought to seek or need more than a fair share, so as to deprive other equally deserving recipients if the wants of the most needy (who may also be the most ill) are satisfied or (2) where an individual's health state poses a danger to others. If anything, in the literature of medical ethics individual recipients of services – that is, patients – are much more likely to be portrayed as endangered by an indifferent or hostile public that denies or delays treatment for them.

Thus, the characterization of recipients of medical services contrasts markedly with the characterization of those who use engineering services. To exercise their autonomy, patients control only the choice of whether to consent or not to the decisions of professionals with knowledge and power of a higher order than their own. Medical ethics even further discounts disabled people's views about their own good, dismissing those who are not inclined to risk whatever functionality they possess to pursue therapies to return them to normality. It is well-known that surveys of people with disabilities find them rating their quality of life higher than expected, but until recently bioethicists have dismissed such reports as products of denial or self-delusion. They have been prompted in doing so by a medical value system that finds it hard to imagine patients enjoying a high quality of life unless they are cured.

Engineering ethics equally has discounted disabled people's views about their own good, but for a different reason. Disabled people have been viewed not as employers or clients, but instead as being subject to the decisions of the real purchasers, namely, physicians or other therapeutic or rehabilitation professionals, family members, or insurance systems. Engineering ethics has tended to adopt these nondisabled people's attitudes toward the disabled and for this reason has not granted primacy to disabled people's self-determination based on their own ideas of the good for themselves.

1.4 The Standard of Normality

Both systems – medical and engineering – presume that the aim of their efforts is to restore the disabled person as closely as possible to normality, understood as species typicality. For medicine, species typical or normal functioning is a component of health [10]. So for medicine, restoration to normal functioning is therapeutic, or at least rehabilitative. Moreover, the closer to normal functioning a person can achieve, the more comfortable the fit with the usual ways of doing things and the less trouble for everyone else who interacts with that person. Returning differently functioning individuals to species typicality therefore may seem to contribute to the general welfare, which aligns with the value engineering ethics places on preserving the public good. Therefore, or at least so it may have seemed, the aims of medical engineering's clients appear consistent with engineers' commitment to the general good.

Adopting species typical functioning as a standard has both benefits and costs. The impaired individual who can be made as good as new escapes the social disadvantages aimed at disability, and society escapes the issues around providing disabled people with special care. Society benefits insofar as one-size-fits-all social arrangements meant only for the species typical may be simpler to organize than practices allowing all to participate. Inclusive practice must be flexible and nuanced in order to respond to people's differences. Further, to the extent that everyone functions alike, resentment from nondisabled people about special accommodations privileging disabled people, and from disabled people about ordinary social arrangements privileging nondisabled people, disappears.

On the other hand, efforts to make disabled people as good as new also have costs, both to society and to the subjects themselves. When normality is the standard that social arrangements expect participants to meet, disabled people feel themselves under great pressure to become as normal as possible, and their families are similarly pressured to get them medical resources so they can be so. Concomitantly, questions of justice about the state's obligation to provide these resources arise. If the state condones public practice that advantages the normal, then should not the state offer those unfortunate enough to depart from normality the opportunity to attain that state?

Yet being offered such opportunity may not be purely beneficial. The same social pressure to be normal may induce people with disabilities to hide their incapacities, even when doing so is dangerous, and to incline them or their families to consent unnecessarily to risky or damaging medical procedures in pursuit of normality. For example, the history of the imposition of dysfunctional prostheses on children with phocomelia is well-known. Among the harms done to these children was the amputation of usable appendages in order to fit more normal looking, but nonfunctional, prostheses [12]. In general, at least some medical interventions aimed at making people more like new fail to do so and also result in reducing their capacity for adaptive functioning.

Further, disabled individuals who fail rehabilitation may suffer loss of self-esteem, viewing themselves as lacking value because they have not reached their long-sought goal of normality. Their trying to do so extracts many kinds of costs, not only psychological damage but also the suffering of educational and career delays because of time spent away from school or work. These considerations all suggest that for many individuals with disabilities the toll taken by engagement with professionals who place too high a value on normality may be unjustly great.

1.5 Justice

Theories of distributive justice are meant to guide decisions about who is owed what resources, and for which purposes, by society as represented by the state. But theories of participatory justice, which determine who counts as a full citizen or subject of justice, come first. In democratic theory, answering this latter question precedes determinations about what kind of political and economic system we should adopt. This is because answers to the questions about the organization of the state, priorities for distribution of resources by the state, citizens' obligations to provide resources to the state, and citizens' entitlements to receive resources from the state all depend on who is thought to merit full considerability in deliberations about the political shape of the state. If we think that moral and political considerability as subjects of justice is deserved only by military males who are native speakers of our language – indeed, who speak it with our provincial accent, as the ancient Greeks did – or only by property-owning white males, as the US once did, then basic principles of justice may be very different than if we have to give equal weight to consideration by very different kinds of people – men and women, native speakers or those who express themselves in a foreign tongue, people of diverse racial heritages and different ethnic identities, and individuals with those physical or intellectual or emotional anomalies labeled "disabilities."

The democratic trend in North America for the past half-century has been to increase the inclusiveness of participation of diverse kinds of people in social and political practices and, by doing so, to improve equitable opportunity for them. Congress and state legislatures in the US (and courts in Canada) are altering how people with disabilities are conceptualized, decreeing that individuals do not become noncitizens just because they function atypically and prohibiting social arrangements that treat them as outcasts, denying them opportunity. The UN Convention on the Rights of People with Disabilities is evidence of how widely this trend has spread. As of this writing, 144 nations have become parties to the Convention.[8]

What does full social participation for people with disabilities involve? First, people's different functional modes must be equally embraced, which means abandoning the standard of normality that privileges typical ways of functioning over unusual ones.[9] While medical professionals aim at restoring patients to functional modes and functional capacity typical of the human species or a familiar subset of humans, engineers prosper professionally by developing a wide repertoire of different functional solutions and offering their clients whichever best suits the situation. Typicality in mode of functioning is not necessarily a value when it comes to designing technology that performs effectively in atypical conditions, and especially in straitened or deprived circumstances.

What follows from this first point is the necessity of civic and commercial organizations flexibly responding to diversely functional people rather than serving only those with species-typical functionality. Building accessibility into technological advances is an especially important requirement of justice. The adverse effect of the telephone's invention on employment of people who are deaf is a well-known example of injustice resulting from technological change. Deaf people were evicted from the workplace by the change from a mainly visual to a mainly aural mode of business communication.

The recent shift back from aural to visual communication, as emailing replaced telephoning and face-to-face speech, threatened a similar result for people who are blind. This was not a technological problem, as the development of voice output of electronic texts followed fairly soon after the means for sending and receiving such texts were widely marketed. There are various accounts of how, in a relatively short time, normal business software was made accessible so that screen reading software works well.[10]

---

[8] See the webpage of the Convention's Secretariat [9] for the latest information on global activity to achieve equitable social opportunity for people with disabilities.

[9] The ethical imperative that promotes this goal is discussed in [11].

[10] For two sides of the story, see [8] on the American Federation of the Blind webpage, and [3] on the Microsoft webpage.

What is clear is that the adoption of state policies prohibiting purchasing of inaccessible programs and the arrival on the market of accessible commercial programs were contiguous. A political commitment to participatory justice prompted just distribution of technological skills and knowledge so that the needed resources were deployed to create products usable by people with different modes of functioning, not just by typically functioning people. Microsoft, for example, now claims "increasing momentum toward the goal of making computers accessible and useful to all people."[11]

1.6 Conclusion

In this chapter, I have drawn attention to an unresolved discordance between medical and engineering values that can conflict the practice of assistive technology. The difference lies in engineering placing a higher value on innovative outcomes, even if these (and perhaps especially if these) alter familiar modes of functionality. I recommend giving engineering values about functionality more preeminence in the practice of assistive technology, and medical values about functionality (at least to the extent that these overvalue "normality") somewhat less.

Assistive technologists have a responsibility beyond designing and distributing prostheses, that is, internal or external devices substituting for or supplementing species-typical physical or cognitive human equipment to enable effective, even if atypical, functioning. Humans are toolmakers, and human society has advanced through the enhancement of functioning that we achieve by using tools. This perspective on people's social participation too often is obscured by romantic prejudices condemning technology as unnatural or unfair.

A common experience for individuals with disabilities when provided with effective assistive technology is to enjoy an enormously expanding vista of opportunity for social participation – from the ability to visit a grocery store at will to the ability to earn a higher education degree – so much so that they feel functionally better than new. But technological progress in enhancing function must be matched by social and political progress in reconceptualizing the relationship between human biology and human technology. Who better than assistive technologists to cultivate public appreciation of the functional intimacy that effectual assistive engineering achieves between disabled people's bodies and their adaptive equipment? Thus, it is not merely permissibly ethical, but professionally obligatory, for assistive technologists to advance justice by promoting progressive ideas and combating cultural prejudices against the prosthetic uses of tools.

---

[11] Compare the narratives at the sites above for two versions of Microsoft's alacrity in pursuing the goal of responsiveness to diversely functioning people. (See footnote 10)

1.7 Afterword

The intersection of the law with medicine in the last decades of the twentieth century reshaped the conceptualization of patients, characterizing receivers of medical care as competent choosers due the respect of deciding their good for themselves. This idea is of individuals who retain the status of full social participants, regardless of their being ill or impaired. The intersection of engineering with medicine has the potential to induce a similar conceptual transformation. Assistive technologists should be in the lead in this movement.

Disabled people who use technology to function come to experience their devices as parts of themselves. Their standpoint in this regard differs from that common to nondisabled people. I illustrate with a personal observation, relating an incident that occurred as I returned to my home city from the site of the workshop that prompted this book.

While I waited with my mobility scooter in the queue to go through US customs, another traveler came up behind me pushing a luggage-loaded large and heavy cart – right into my scooter. As I swung my seat around on the now immobilized scooter to see a back wheel severed from the axle and lying on its side, the perpetrator laughingly said that she had been looking back over her shoulder for a friend and so had not noticed me, and then proceeded on to her plane – leaving me to deal with mobility as impaired by her irresponsibility as if she had broken my leg, and wondering why the workshop I had just attended had facilitated discussion about the ethical obligation of protecting walking people from injury by wheelchair users, but not about the equally harmful reverse.

Justice for all equally calls for affording engineering's emendations to human functioning the same status as medical interventions. One is as human an enterprise as the other, for disabled people who use assistive technology are as human as nondisabled people who do not.

References

[1] Beauchamp T, Childress J (2008) Principles of Biomedical Ethics. Oxford University Press, New York
[2] Cooper R (2006) Wheelchair standards: It's all about quality assurance and evidence-based practice. Journal of Spinal Cord Medicine 29(2): 93–94, http://www.ncbi.nlm.nih.gov/pmc/articles/PMC1929010
[3] Microsoft Accessibility (2010) History of Microsoft's commitment to accessibility, http://www.microsoft.com/enable/microsoft/history.aspx
[4] National Academy of Engineering (2003) Online ethics center for engineering and research, http://www.onlineethics.org/

[5] Rauhala M, Topo P (2003) Independent living, technology, and ethics. Technology and Disability 15(3):205–214

[6] Rehabilitation Engineering and Assistive Technology Society of North America (2010) Code of ethics, http://resna.org/certifications/certification-professional-standards-board

[7] Rhodes R, Francis L, Silvers A (eds) (2007) The Blackwell Guide to Medical Ethics. Wiley-Blackwell, Oxford

[8] Schroeder P (2000) A brief history of Microsoft and accessibility. AccessWorld 1(4), http://www.afb.org/AFBPress/pub.asp?DocID=aw010402

[9] Secretariat for the Convention on the Rights of Persons with Disabilities (2008) Promoting the rights of persons with disabilities, http://www.un.org/disabilities

[10] Silvers A (2001) No basis for justice: Equal opportunity, normal functioning, and the distribution of healthcare. American Journal of Bioethics 1(2):35–36

[11] Silvers A (2001) Prescribing multi-functionalism to achieve equality in a world of difference. Health Ethics Today 12(1), http://www.phen.ab.ca/materials/het/het12-01a.asp

[12] Silvers A (2003) On the possibility and desirability of constructing a neutral conception of disability. Theoretical Medicine and Bioethics 24(6):471–487

[13] Silvers A (2008) The right not to be normal is the essence of freedom. Journal of Evolution and Technology 18(1):79–85, http://jetpress.org/v18/silvers.htm

[14] Steinbock B, London AJ, Arras J (eds) (2008) Ethical Issues in Modern Medicine: Contemporary Readings in Bioethics. McGraw-Hill, New York

[15] Vaughn, Chairperson J (2006) Over the horizon: Potential impact of emerging trends in information and communication technology on disability policy and practice. Washington, DC, http://www.ncd.gov/newsroom/publications/2006/emerging_trends.htm

[16] Williams CJ (2010) Blind UCLA graduate can use computer-assisted reading tools during state bar exam. Los Angeles Times, http://articles.latimes.com/2010/feb/05/local/la-me-blind-bar6-2010feb06

[17] Woolston C (2010) The healthy skeptic: Be wary of products touting FDA certification. Los Angeles Times p 1, http://articles.latimes.com/2010/mar/01/health/la-he-0301-skeptic-20100301, Health section, Features desk, Part E

Chapter 2
The Myth of Neutral Technology

Jan Miller Polgar

Abstract The meaning that assistive technology (AT) holds for the user is a key determinant of whether the device will be used or abandoned. Two concepts, stigma and liminality (existing in a state of transition), are used to frame users' perceptions of the assistive technology they use, as generated through research projects investigating aspects of assistive technology use. Implications of the meaning of AT to the design and selection process are described. Assistive technology that is seen as a tool, as just another way of achieving a desired activity is much more likely to be assimilated into the user's daily life. Technology perceived in this manner enables people to share activities with others and augment their personal abilities. Alternately, technology can be seen as a visible sign of disability, reinforcing stigma associated with a disability and the perception of the AT user as existing somewhere between health and illness. Individuals with this view of technology may avoid or resist use of technology, resulting in avoidance of meaningful activities and both social and physical isolation. These findings support the conclusion that technology is not neutral. Inclusion of users in both the design and selection process and understanding the meaning that AT use holds are integral to the development of assistive technology that achieves the desired outcome of enabling participation in daily life.

> Meaning is at least as powerful an influence as skill in determining whether a device will be incorporated by an individual as a useful tool or discarded as excess baggage [9].

Assistive technology (AT) can augment or replace function in many individuals with disabilities, enabling them to participate in daily activities in their communities. The functionality of AT can be very appealing to

Jan Miller Polgar
School of Occupational Therapy, University of Western Ontario, London, ON, Canada,
jpolgar@uwo.ca

M.M.K. Oishi et al. (eds.), *Design and Use of Assistive Technology: Social, Technical, Ethical, and Economic Challenges*, DOI 10.1007/978-1-4419-7031-2_2,
© Springer Science+Business Media, LLC 2010

designers, researchers and prescribers, with the potential to create ever more complex technology to address physical, cognitive, or sensory impairment without the necessary consideration of the influence of technology on the user's self-perception. Focus on device functionality may limit understanding that the technology holds meaning to the consumer and significant others, in other words, that technology is not neutral. Meaning is often a factor in whether technology will be used or put into the closet. This chapter will explore various meanings that users ascribe to AT and the implications of these meanings to the design, recommendation, selection and evaluation processes. Two related constructs will guide this exploration: stigma and liminality.

Stigma was described by Goffman [2] as possession of an external characteristic that discredits the individual. The presence of a physical impairment becomes the source of a spoiled identity. He discusses the concept of "spread" where the discrediting attribute attains a "master" status so that it defines the individual, and all other personal accomplishments or attributes are ignored. Stigma is context dependent; environments of various physical, social, or institutional elements either reinforce or limit the perception of stigma.

Liminality is an anthropological term that conveys the notion of transition and is often used to characterize the period of development of moving from childhood to adolescence. It frequently involves a change of status, social isolation, and/or physical removal of an individual. Liminality for a person with disabilities has been described by Murphy [7] as feeling distant from society, although not specifically excluded from it; as being between health and illness. Murphy's 1990 book *The Body Silent* describes his transition from full participation in society as an academic anthropologist to living with the physical abilities that resulted from a spinal cord tumor that caused increasing paralysis. His account ably describes the experience of being and becoming a person who has a disability [10].

These two constructs, stigma and liminality, are used to frame the discussion of the meaning of assistive technology for individuals who use it, including the consumer and their families or other caregivers. They will be applied to the following ideas to give context to results of various research projects that included a component of the meaning that individuals ascribe to the technology they use and to promote awareness in AT designers and prescribers of why a device that is anticipated to be of benefit to the user in their daily life is not embraced or is even discarded.

2.1 Source of the Data

The following ideas come from various research projects: a qualitative exploration of how persons with disabilities chose to complete daily activities, a phenomenological study of the lived experience of using AT when assisting a significant other, and design projects that involved consumers (broadly understood) at all stages of the process as well as from the literature [6, 5]. This work, in part, derives from concern about the rate of abandonment of AT and the associated costs – monetary, time, and social participation – to the consumer, others in their social sphere, and society.

2.2 Key Ideas Related to Meaning and Assistive Technology

Not surprisingly, the meaning of technology conveys both positive and negative/enabler and barrier notions. Two broad perceptions of technology will be discussed to illustrate the meaning that AT can hold for consumers: *assistive technology as a tool* and *assistive technology as a visible sign of disability*.

2.2.1 Assistive Technology as a Tool

Some assistive technology users, mostly those who are experienced users, view their technology as a means to complete an activity. The meaning ascribed to the AT was no different than any other tool or technology used in daily life. For example, participants in a project exploring how individuals chose to complete daily activities talked about sharing opinions on the best chair components for their sport along with wheelchair maintenance and modification activities. These discussions had the flavor of cyclists discussing the latest cycling technologies. The AT was simply a tool that enabled them to compete in a sport they loved.

Participants in both qualitative studies indicated that assistive technology facilitated their ability to engage in desired activities. Participants in one study indicated that access to AT was a primary factor in enabling a significant other to return home, as seen in the following quote [6]:

> "[lift] helps him and me together to do it [daily care]... Without [the lift] I would not be able to handle [husband]."

For many participants in these studies, the technology had faded into the background of their lives. They did not speak of the challenges posed by the technology but did describe the constraints imposed by the environment that was not accessible due to their use of the technology.

Assistive technology can provide a means to share activities. While the decision to obtain a powered wheelchair is not made lightly, it can provide independent mobility that results in a couple being able to walk side by side as they go about daily tasks rather than one partner being dependent on the other to propel the chair. Technology levels the playing field, a point often made in reference to computer access technology. As one study participant indicated [5]:

> "We are all given the exact same abilities in the digital environment ... And when you are online, nobody knows you have a disability so it [the disability] never really comes into it—it has a very big appeal to me."

When technology is viewed as a tool, as something that is a necessary and integrated part of daily life, it becomes an enabler of activity. However, far too often, technology is seen as a reminder that the user cannot participate in their community as they wish.

2.2.2 Assistive Technology as a Visible Sign of Disability

When assistive technology is perceived to signal to others that the user is disabled, it can become a barrier to participation in daily activities or be abandoned. The technology reinforces a discrediting attribute and enhances the perception of stigma. The technology becomes the focus of attention rather than the person using the technology. The influence of the technology on others' perceptions is seen in the following quote by a 22-year-old university student [5]:

> "People see the chair and make certain perceptions and there's certain understanding of what it means to be disabled."

Some people will avoid the use of technology and either not go to certain community locations or, if they do so, will limit what they do because they do not want to be seen as someone with a disability. The following quote from a 25-year-old social worker illustrates this point [5]:

> "the other night...I chose to like suck it up and walk the best I could without my cane, because I would rather them not see me like that."

Similar ideas were expressed by people who refused to use a white cane or a powered wheelchair because of the image of disability these devices conveyed.

AT can be seen as a source of isolation in a number of ways. It can physically isolate the user as in the experience of a woman who reported that she and her husband were located in a back corner of a restaurant, out of sight of other patrons or when people who use wheelchairs are relegated to specific areas of a theatre or stadium that are accessible, but that do not necessarily allow them to sit with their companions. Technology such as augmentative and alternative communication devices or wheelchairs, by the nature of their design and use, are an actual physical barrier between the user and others in their environment, reinforcing a state of liminality.

More troubling than the physical barrier is that AT can be socially isolating. A common comment, reflected in the notion of wheelchairs as portable seclusion huts [8], is that the technology distances the user from others in their environment. This physical barrier then becomes a psychological barrier. People who use technology frequently report experiences such as service providers speaking to their companions rather than themselves in social circumstances, such as when they are making purchases.

2.3 Implications for AT Design and Selection

The collective results of these projects reinforce the notion that assistive technology is not neutral. It carries a meaning to the user, others in their social sphere and the community at large. When that meaning is a positive one, AT is more likely to be incorporated into daily life. These key themes and the feedback obtained during the design process have important implications for AT design and selection. They suggest that technology that reduces the impact of stigma and that enables the user to participate and thus feel less isolated will be more acceptable and useful to the consumer.

Successful technology fades into the background of daily life. It becomes just another way of completing a desired activity. Feedback provided by children, parents and formal caregivers on design and function of AT in the development of three types of seating projects [1, 4] revealed considerable consistency across projects on perceptions of features that defined a desirable device. Many consumers wanted a device that was aesthetically pleasing, was simple to use and easy to maintain. Flexibility of use for different activities and in different environments was another key feature.

The design and appearance of the device carry meaning for the individual and should not be overlooked in the development or recommendation processes. Color of the device may not seem like an important feature over others such as the ability to tilt a wheelchair seat or adjust a guardrail on a bed. However, materials used to make a device and attention paid to the aesthetics can result in a device that blends into the user's environment, or not. A device that looks similar to those available on the consumer market reinforces the perception that the user is healthy and potentially reduces isolation (ideas related to liminality).

The meaning of technology is a particularly important consideration for older adults, who form an increasingly larger proportion of our population. Many older adults want to remain in their own homes and will incorporate technology that does not make them feel vulnerable and allows them to feel more secure, safe and less of a perceived burden. Johnson, Davenport and Mann [3] explored the perceptions of older adults of different smart technology used in the Gator House at University of Florida. They asked seniors whether they felt that the devices present in the house were useful to them at this time in their life. Technology that conveyed a sense of security was more readily accepted than that which was seen as surveillance. For example, people typically indicated that sensors that would detect a fall were probably something that would be useful but not at that point in their life; in other words, the more invasive technology was less acceptable.

The collective results of all of these projects lead to one critical conclusion: the individuals seeking devices are the experts in their own lives. The individual designing or recommending a device retains expertise related to AT and a professional responsibility to the client, but does not know what is best for the user. It remains important to ask questions about what technology means to the user. The most carefully designed and prescribed technology is of no value if the intended user leaves it in the closet. Technology that is recommended without input from the consumer is in danger of being abandoned, at a cost to the user, their community, and society.

References

[1] Fong Lee D, Ryan S, Polgar J, Leibel G (2002) Consumer based approaches used in the development of an adaptive toileting system for children with positioning problems. Physical and Occupational Therapy in Pediatrics 22(1):5–24

[2] Goffman E (1963) Stigma: Notes on the management of spoiled identity. Simon and Schuster, New York

[3] Johnson J, Davenport R, Mann W (2007) Consumer feedback on smart home applications. Topics in Geriatric Rehabilitation 23(1):60–72

[4] Miller Polgar J, Ryan S, Coiffe M, Barber A (2003) Development of a toileting system for adolescents with severe positioning problems: Feedback from consumers. In: 26th International Conference on Technology and Disability: Research, Design, Practice and Policy (RESNA), Atlanta, GA

[5] Miller Polgar J, Winter S, Howard S, Maheux K, Nunn J (2009) The meaning of assistive technology use. In: Proceedings of the 25th International Seating Symposium, Orlando, FL, p 75

[6] Morgan-Webb S (2005) Understanding the informal caregiver's experience with the use of assistive devices. Master's thesis, Occupational Therapy, University of Western Ontario

[7] Murphy R (1990) The body silent. W. W. Norton, New York

[8] Scheer J (1984) They act like it was contagious. In: Hey S, Kiger G, Seiden J (eds) Social aspects of chronic illness, impairment, and disability, Willamette University, Salem, OR

[9] Spencer C (1998) Tools or baggage? Alternative meanings to assistive technology. In: Gray DB, Quantrano L, Lieberman ML (eds) Designing and using assistive technology: The human perspective, vol 2, Paul H. Brookes, MD, pp 89–98

[10] Wilcock A (2006) An occupational perspective on health, 2nd edn. SLACK, Thorofare, NJ

Chapter 3
Accessible Technology and Models of Disability

Richard E. Ladner

Abstract In this chapter, we discuss assistive technology from the view of the consumer. Consumers of assistive technology follow the social model of disability, that is, persons with disabilities are part of the diversity of life, not necessarily in need of cure or special assistance. Their identity does not revolve around being a patient or client, but focuses on their human desires to work, play, and associate with others. The social model of disability dictates an empowering approach to assistive technology research and development where consumers are given the power to configure and even create technology to suit their own needs and desires. The technology that comes from this approach is called accessible technology, rather than assistive technology, emphasizing its role in making human activities more accessible.

3.1 What is Assistive Technology

"Assistive technology" is really a redundant term because, in some sense, all technology is assistive, making tasks possible or easier to do. Automobiles, trains, busses, and airplanes are assistive technologies because they make getting from one location to another easier to do. Telephones assist us to talk to people over a distance. Computers are very general devices that assist us with many tasks. In some sense, the whole purpose of technology is to make tasks possible or easier to do. Nonetheless, the moniker "assistive technology" has come to mean specialized technology for persons with disabilities. Hearing aids are commonly called assistive listening devices, yet eye-glasses and contact lenses are not typically called assistive technology. Correctable vision is not considered to be a disability. Why is it that persons with disabilities have assistive technology, while the rest of us just have technology?

Richard E. Ladner
Computer Science, University of Washington, Seattle, WA, ladner@cs.uw.edu

M.M.K. Oishi et al. (eds.), *Design and Use of Assistive Technology: Social, Technical, Ethical, and Economic Challenges*, DOI 10.1007/978-1-4419-7031-2_3,
© Springer Science+Business Media, LLC 2010

It is helpful to see what terms are used by advocacy groups of persons with disabilities. The National Federation of the Blind just uses the term "technology" on its web site for technology that is useful to its members, such as screen readers, Braille printers, Braille notetakers, and talking phones. Similarly, the American Council of the Blind talks about "products," not "assistive technology products". The National Association of the Deaf just uses the term "technology" on its web site for technology that is useful to its members such as captioning, video phones, and video relay services. However, they do use "assistive listening technology" for devices that enhance hearing. The Alexander Graham Bell Association for Deaf and Hard of Hearing talks about hearing aids and cochlear implants, but does not use the term "assistive technology" on its web site. Why do many people with disabilities not use the term "assistive technology" when talking about the technology that they use?

I believe that the two questions in the previous paragraphs are answered by understanding how people with disabilities view themselves and their relationship to technology; that is, their identity as users of technology. First, as I pointed out earlier, the addition of "assistive" does not add anything to "technology" that one uses every day on a routine basis. All technology is assistive, by definition. Second, and more important, the term "assistive" when used with "technology" emphasizes a person's need for extra assistance. It has the ring of paternalism, a view that people with disabilities need lots of extra help, are dependent and are not capable human beings.

3.2 Models of Disability

There have been a number of attempts to define different perspectives on or models of disability [1]. The following are five models of disability that are found in western society through which we can understand better the relationship between people with disabilities and technology.

Medical Model: People with disabilities are patients who need treatment and cure or partial cure. Generally, treatment is very expensive and may require continual monitoring over a lifetime. Some assistive technologies can be prescribed by doctors or other medical professionals. In such cases, the technologies may be paid for by medical insurance.

Rehabilitation Model: People with disabilities are clients who need assistance and assistive technology for employment and everyday life. Job related assistive technology may be paid for by employers, but technology for everyday living is usually paid for by the client.

Special Education Model: Children with disabilities have the need for special education, which may differ substantially from education programs that other children receive. Technology may be provided to some children with disabilities to further their education. In the United States the Individuals with Disabilities Education Act (IDEA) provides for access to public education for all children regardless of disability. Under this act a student may have an Individualized Education Plan (IEP) whereby the student may not be expected to achieve as much as other children without a disability.

Legal Model: People with disabilities are citizens who have rights and responsibilities like other citizens. Accessibility to public buildings and spaces, voting, computers, television, and telephone are some of those rights. These rights are ensured by many laws in the United States including the Rehabilitation Act and the Americans with Disabilities Act (ADA). Often assistive technology related to legal access is provided free of charge.

Social Model: People with disabilities are part of the diversity of life, not necessarily in need of treatment, cure, or special assistance. They do need access – often through technology – to partake in many activities of life, but they do not need to be taken care of or have decisions made for them by others. Technology developed in the social model is typically paid for by the individual with the disability or is free, if mandated by law. Closed captioned television is an example of free technology mandated by law.

Naturally, these five descriptions of models of disability are oversimplified, but they do help delineate overlapping approaches to disability. Each of the models has a work force that provides the services and products associated with the model. Medical professionals provide medical procedures and services. Rehabilitation professionals provide services like job training and assistive technology training. Special education professionals teach in the classroom and provide out-of-class training. Disability lawyers and other legal professionals provide legal assistance when needed, for example, to file a law suit. The social model workforce includes developers and sustainers of technology whose goal is technology at no or low cost to consumers. There are also certain workforces that cut across all five models. For example, sign language interpreters and real-time captioners are needed by deaf people in many situations including medical, educational, work place, legal, and social.

3.3 Accessible Technology

These models of disability provide a framework for thinking about the relationship between people with disabilities and the technology they use. The medical model tends to focus on prostheses, devices that restore lost function. For example, the cochlear implant is a surgically implanted device that stimulates the auditory nerve system and may partially restore hearing. Joint replacement surgery may restore someone's ability to walk. Research on retinal implants may partially restore vision someday. The rehabilitation model tends to focus on assistive technology that does not necessarily restore lost function, but permits alternative approaches to achieving goals. For example, a motorized wheelchair does not allow someone to walk, but provides an alternative to walking for someone desiring to go from one location to another. A video phone does not restore hearing, but instead allows a deaf person to talk at a distance to another person in sign language.

In the social model, the term "accessible technology" seems to fit better than "assistive technology" because it focuses on alternative approaches to achieving goals rather than the paternalistic notion of needing assistance. Given the prevalence of the medical and rehabilitation models one should not expect a sudden change in language use. The term "assistive technology" will continue to be used for many years to come. However, for those of us working in accessible computing research – which is not currently dominated by the medical and rehabilitation point of view – the term "accessible technology" is very attractive. The term leans towards the legal and social models, and away from the paternalistic medical and rehabilitation models. The term "assistive technology" has come to mean special purpose devices, but given that computers are general purpose devices that can provide multiple accessibility solutions the term "accessible technology" might be more appropriate.

3.4 Concepts from Human–Computer Interaction

Human–computer interaction (HCI) is one research area where there is significant interest in accessible technology. At the premier HCI conference, the International Conference on Human Factors in Computing Systems (CHI), 32/277 (12%) papers mentioned the term "disability" in 2009. The ACM SIGACCESS Conference on Computers and Accessibility (ASSETS) started in 1994 and focuses exclusively on accessible technology. The interest among HCI researchers in accessibility solutions for people with disabilities is rooted in the researchers' innate desire to solve interesting problems that are related to computers and human beings. Some of the most interesting problems come from the space where the humans have some disability.

One concept from HCI is *usability*, meaning how easy it is to learn, configure, and use an interface. It is possible for an interface to be accessible but not especially usable. Examples are early audio CAPTCHAs[1] that permit blind users to prove that they are human by listening to a distorted message and then entering the message into a text box. These audio CAPTCHAs allowed blind users to access sites that only permit human access, but they are extremely difficult to use. Another important concept in HCI is *human centered design*, which means involving users in each iterative design cycle from concept to prototype to final design. Human centered design generally leads to

---

[1] A CAPTCHA is a challenge response test often used by web site servers to determine whether the entity trying to access the site is a human or another computer. The most commonly used CAPTCHAs are images of distorted characters.

more usable interfaces. The same principle can be applied to the design of accessible technology by involving disabled users in the design process. The concept of *universal design* means designing interfaces in a way that they can be used by the vast majority of people regardless of disability. For webpage design this might mean making sure that the design is usable by a blind person with a screen reader, without necessarily modifying its visual appearance. Fortunately, there are standards for such designs in the web content accessibility guidelines (WCAG) offered by the W3C [6].

It is doubtful that one interface can be truly universal, so another approach to accessibility is to *design for user empowerment*, which means to design to enable people to solve their own accessibility problems whenever possible. A simple example is found in modern screen readers where the user can adjust the speed of the speech. Many blind people use a very high speed which is incomprehensible to an average listener. In screen readers, a design goal is not the most natural speech, but speech that can be understood at a speed adjustable by the user. Designing for user empowerment is a non-paternalistic approach to interface design. In a sense, it encompasses both human centered design and universal design. It would be difficult to design for user empowerment without involving users in the design cycle. Universal design includes solutions that allow for easily set individual configurations; for example, most computer keyboard trays have adjustable height.

The most powerful form of user empowerment is providing the education and environment to people that will enable them to solve their own accessibility problems. The famous adage, "Give a man a fish; you have fed him for today. Teach a man to fish; and you have fed him for a lifetime," applies to accessible technology. The highest level of user empowerment is the inclusion of people with disabilities as designers and creators of the technology.

3.5 New Directions in Accessible Technology

There are many new research directions in accessible technology that take the non-paternalistic approach of the social model of disability, and in this last section I will mention a few. The first is the SUPPLE++ project, which automatically generates appropriate user interfaces for users with varying motor and vision abilities [4]. The beauty of this system is how easy it is for a user to configure the interface: A user can perform a few simple tasks, which are recorded and analyzed to determine which interface would work best for that person.

The second is WebAnywhere, an open source web-based screen reader [2]. WebAnywhere is a free screen reading web service that allows audio access to the web from almost any computer, even those with highly circumscribed capabilities such as those found in public places like libraries and internet cafes. Because it is open source, contributors from around the world, disabled or not, can add value to the service by adding new features, such as multiple languages. The development of open source software for accessibility fits the social model of disability.

The third is the ASL-STEM Forum, which enables people to upload definitions and their signs for topics in science, technology, engineering, and mathematics (STEM) [3]. Because deaf students are so thinly spread at hundreds of colleges and universities it is very difficult to develop a uniform vocabulary of signs in technical fields. The forum will enable students, their sign language interpreters, and deaf professionals to share vocabulary and discuss among themselves what signs they would like to use. The forum is a form of community empowerment whereby deaf students can solve their own accessibility problems collectively.

Finally, I will mention the MobileAccessibility Project [5], which is a relatively new project at the University of Washington and University of Rochester. Modern smart phones have multiple sensors (camera, microphone, GPS, compass, accelerometer), network capability (WiFi, Bluetooth, cellular data network, cellular voice network), alternative input (keyboard, touchpad, buttons, speech), and output (screen, sound, speech, vibration). With appropriate programming these devices could solve multiple accessibility problems, that is, smart phones can become multi-function accessibility devices. As an example, smart phones (KNFB Reader Mobile, TestScout) can already do optical character recognition (OCR) so that blind people can take pictures of pages of text, convert them to digital text, and then listen to them using text-to-speech technology. My vision for the future is that a blind person can buy a standard smart phone and data plan, and then download any needed accessibility applications. Blind designers and computer programmers will be part of the development community that builds these new applications.

3.6 Conclusion

People with disabilities, including those who become disabled as a result of aging, are all part of the rich fabric of human life. Modern medicine's ability to save lives creates even more people with disabilities before old age sets in. Typically, people with long-term disabilities are not looking for cures or leading lives of desperation. People who are newly disabled go through a number of stages of grief that eventually turn into some form of acceptance of their disabilities. In either case, people with disabilities are simply social human beings who want to work, play, and associate with friends. Technology enables these people to have more access to what they want and need in life. Designing technology for user empowerment fits this social model of disability. "Accessible technology" is a better term than "assistive technology" for technology targeted to people with disabilities who view themselves in the social model of disability.

Acknowledgment Thanks to Anna Cavender for several helpful comments on a draft of this chapter.

References

[1] Albrecht GL, Bury M, Seelman KD (ed) (2001) Handbook of Disability Studies. Sage publications, California

[2] Bigham JP, Prince CM, Ladner RE (2008) WebAnywhere: a screen reader on-the-go. In: Proceedings of the international cross-disciplinary Conference on Web Accessibility (W4A 2008), pp 73–82, http://webanywhere.cs.washington.edu

[3] Cavender A, Otero D, Bigham JP, Ladner RE (2010) ASL-STEM Forum: Enabling Sign Language to Grow Through Online Collaboration. In: 28th ACM Conference on Human Factors in Computing Systems (CHI 2010), Atlanta, GA, pp 2075–2078, http://aslstem.cs.washington.edu

[4] Gajos KZ, Wobbrock JO, Weld DS (2007) Automatically generating user interfaces adapted to users' motor and vision capabilities. In: 20th Annual ACM symposium on User Interface Software and Technology (UIST 2007), pp 231–240, http://www.eecs.harvard.edu/~kgajos/research/supple/

[5] MobileAccessibility (2008) Bridge to the world for blind, low-vision, and deaf-blind people. Computer Science and Engineering, University of Washington, Seattle, http://mobileaccessibility.cs.washington.edu

[6] W3C (2008) Web Content Accessibility Guidelines (WCAG) 2.0, http://www.w3.org/TR/WCAG20/

Chapter 4
The Importance of Play:
AT for Children with Disabilities

Albert M. Cook and Kim Adams

Abstract The potential of robots as assistive tools for play activities has been demonstrated through a number of studies. Children with motor impairments can use robots to manipulate objects and engage in play in activities that parallel those of their typically developing peers. This participation creates opportunities to learn cognitive, social, motor, and linguistic skills. By comparing disabled children's performance with that of typically developing children, robot use can also provide a proxy measure of cognitive abilities.

4.1 Background

During typical development, play activities provide an opportunity for children to learn cognitive, social, motor, and linguistic skills through the manipulation of objects. Physical impairments make it difficult for a child who has disabilities to independently manipulate objects in a play context [16].

Assistive technologies have been used to enable play by adapting battery powered toys to be controlled by a single switch activated by a gross motor movement. This engages children and provides a sense of control, but the repetitive action of the toy causes the child to lose interest. There are also simple electronic aids to daily living (EADLs) that allow an appliance such as a food mixer to be plugged in and controlled with a single gross movement on one switch. This allows a child to participate in activities with other children. For example a child with fair fine motor control could open a

Albert M. Cook
Speech Pathology and Audiology, Faculty of Rehabilitation Medicine,
University of Alberta, AB, Canada, al.cook@ualberta.ca

Kim Adams
Glenrose Rehabilitation Hospital, and Faculty of Rehabilitation Medicine,
University of Alberta,
AB, Canada, kim.adams@ualberta.ca

M.M.K. Oishi et al. (eds.), *Design and Use of Assistive Technology: Social,*
Technical, Ethical, and Economic Challenges, DOI 10.1007/978-1-4419-7031-2_4,
© Springer Science+Business Media, LLC 2010

package of pudding and pour it into a blender. Another child might add milk and a child with very limited motor control could mix the ingredients using the EADL.

One challenge with many EADL or switch-controlled toy situations is that a child who is using augmentative communication must choose between controlling the toy or EADL, or controlling her communication device. Anderson [4] overcame this choice problem by having children control infrared toys from their communication device and reported that this approach offers "highly motivating activities for use in the development of language" (p. 7). Although this approach solves the problem of integrating play and communication, infrared toys will always perform the same function, becoming boring after a while.

4.2 Robot Applications for Children

Robots have a potential learning advantage over toys or appliances since robots can be reprogrammed to perform a variety of functions and thus keep the interest of the child. They can also present increasing challenges. Young children with disabilities can control robots for manipulation of three-dimensional objects in play and school activities. Children with disabilities may also be able to carry out the robot programming on a computer to engage in problem solving.

Robots have been used successfully in a number of studies to allow children with disabilities to participate in play and engage in school-based activities. School activities aided by robots include manipulative tasks using a robot [12], pick and place academic activities [15, 14], drawing [21], science lab activities [11], and science and art [10, 17]. Robots have also been used to facilitate play [5, 13]. In many of these projects, the focus is on the importance of play in child development and the ability of robotics to enable play by children who have disabilities. The IROMEC project team has developed a set of play scenarios that serve to set the context for users to be involved in the design process of appropriate robotics activities and hardware. They have identified four types of play: sensory motor play, symbolic play, constructive play, and games with rules [20], and have developed a flexible modular mobile robot to accommodate multiple users and play scenarios [18].

4.3 Robots and Cognitive Development in Children

Many standardized tests of cognitive ability require speech or fine motor control, which can make it difficult to ascertain the developmental level of children who have motor disorders and complex communication needs. Due to

Fig. 4.1 Lego robots and adapted controller

these limitations, children with severe disabilities may be perceived as being more developmentally delayed than they actually are. Robots give children an opportunity to manipulate items and choose how to interact with their environment. Because these tasks often require problem solving, they can also provide a method for children to demonstrate their understanding of cognitive concepts.

Robots have been used to demonstrate previously unmeasured cognitive skills. Robot-based tool use was demonstrated by disabled and typically-developing children greater than 8 months in mental age by using a robot to bring an object closer to them [6]. Children aged 6–14 years who had severe cerebral palsy performed a structured play task to uncover a hidden toy by activating one or more switches [7]. The majority of the participants could not be evaluated through standard cognitive measures, but teachers noticed differences in overall responsiveness, amount of vocalization, and interest (i.e., increased attention to tasks) for children who used the robotic arm.

Ten children with varying physical and cognitive disabilities participated in a study using the Lego MindStorms robot [8]. Two robots were built and used in this project, and are shown in Fig. 4.1: the robot arm (left image in the figure) and the roverbot car (centre image). Each robot could be programmed to perform different actions. The child used from one to four switches connected to a modified remote control (right image) designed and built for this project. The child could play back a movement (e.g., a dancing robot) with one switch press or control the robot to move in four directions (left, right, forward, back) using four switches.

The study's hypothesis was that children with cognitive disabilities will use a robot to interact with objects in a manner that is consistent with typical developmental levels for non-disabled children.

The children fell into three groups with respect to their cognitive level and skills attained with the robots:

1. *Severe physical and cognitive disabilities:* These children had limited control or understanding of the robot. They controlled the robot with a single switch to bring toys to them, or to move it across their field of vision.

2. *Multiple physical and cognitive disabilities:* This group had greater, but still limited, control over their surrounding environment. They were able to use one switch to control the robot in tasks such as fetching a toy or taking toys to play partners. They took turns with the researcher. This group's verbal skills improved, their willingness to interact with others increased, and their ability to concentrate on new tasks was apparent throughout the sessions.

3. *Greater physical and cognitive abilities:* These children controlled the robots using multiple switches. This group could drive the robot through an obstacle course, create stories, use the robot to take specific items to others, and use the arm to sort and play games with other students. Their socialization skills increased, and they became more outgoing and vocal. Their parents were pleased as they noticed changes in the home environment.

Children in groups 2 and 3 demonstrated discovery using the robots in symbolic and imaginative play. All of the children were able to demonstrate a range of cognitive skills, even though many of them were judged non-testable on standardized tests. The hypothesis was proven true, and a hierarchy of cognitive skills represented by robot tasks was developed.

4.4 Robot Use by Very Young Typically Developing Children

If robot-based tasks are to be used as a proxy measure of cognitive development, it is important to know how typically developing young children are able to use the robot. Few studies have addressed robot use with very young children. A recent study involved typically developing children aged 3, 4 and 5 years [19]. They used a Lego robot to complete tasks based on the cognitive concepts of causality, negation, binary logic, and sequencing. All of these tasks are related to the use of electronic assistive technologies and the use of robots for exploration and discovery.

Of the cognitive skills, causality was understood by all of the participants, negation (the concept that releasing a switch was an action) and binary logic (left and right) task were understood by the 4-and 5-year-olds. The 3-year-olds had more difficulty with negation and none were able to consistently use a two-step sequence to accomplish a task. Most of the 4-and 5-year-olds accomplished the two-step sequence successfully. This study confirmed that robot-related tasks were dependent on developmental level. This provides the basis for using simple robot tasks to probe cognitive understanding and developmental level in children who have disabilities.

4.5 Integrating Communication and Robotic Manipulation

As described above, children who use augmentative communication devices (often called speech generating devices or SGDs) may have difficulty in integrating the control of play objects with the control of the SGD. Just as Anderson [4] integrated SGDs and infrared controlled toys, infrared robots like the Lego roverbot can be controlled via the SGD [3]. Many SGDs have the capability to learn infrared commands. Robotic control is important because much of play and selected portions of the academic curriculum involve manipulation of real objects. Controlling robots through SGDs allows children to talk while they play, similar to how typically developing children do it.

Professional experts [9] and children with and without disabilities performed usability testing with an integrated communication and robotic play platform. The robot could be used to either play back stored programs or in direct teleoperated mode. Teleoperation was possible for the experts and older children (5-year-olds). Younger children (3-year-olds) relied on the playback of pre-programmed movement using a single switch. Children were also given the option of directing another person to do the manipulation in a play task or to do it directly with the robot. Children preferred to do activities using the robot rather than directing another person. All children used the built-in communication functions to spontaneously talk while using the system during play.

A commercially available communication device was used by a 12-year-old girl who has Cerebral Palsy to control Lego robots for academic activities [3, 2, 1]. This study established the feasibility of controlling Lego robots via an SGD for social studies, math, and robot programming activities. With systems such as these, children can demonstrate and develop manipulative, communicative, and cognitive skills in an integrated way.

4.6 Summary

Small robots can provide interesting and engaging opportunities for children with disabilities to engage in play. They can also allow access to learning activities involving manipulation. The ways in which children use robots reveals a great deal about their cognitive skills and problem solving abilities. When combined with use of augmentative communication devices, robots provide a powerful active component to play and academic activities that is not possible with the communication device alone.

References

[1] Adams K, Cook A (2009) Using an augmentative and alternative communication device to program and control Lego robots. In: RESNA Annual Conference, New Orleans, LA

[2] Adams K, Yantha J, Cook A (2008) Lego robot control via a speech generating communication device for operational and communicative goals. In: International Society for Augmentative and Alternative Communication – 13th Biennial ISAAC Conference, Montreal, QC

[3] Adams K, Yantha J, Cook A (2008) Lego robot control via a speech generating communication device for play and educational activities. In: RESNA Annual Conference, Washington, DC

[4] Anderson A (2002) Learning language using infrared toys. In: 23rd Annual Southeast Augmentative Communication, Birmingham, AL

[5] Andreopoulos A, Tsotsos J (2007) A framework for door localization and door opening using a robotic wheelchair for people living with mobility impairments. In: Proceedings of the Robotics Science and Systems (RSS) Manipulation Workshop: Sensing and Adapting to the Real World, Atlanta, GA

[6] Cook A, Hoseit P, Ka M, Lee R, Zenteno-Sanchez C (1998) Using a robotic arm system to facilitate learning in very young disabled children. IEEE Transactions on Biomedical Engineering 32(2):132–137

[7] Cook A, Bentz B, Harbottle N, Lynch C, Miller B (2005) School-based use of a robotic arm system by children with disabilities. IEEE Transactions on Neural Systems and Rehabilitation Engineering 13(4):452–460

[8] Cook A, Adams K, Volden J, Harbottle N, Harbottle C (2010) Using Lego robots to estimate cognitive ability in children who have severe physical disabilities. Disability and Rehabilitation: Assistive Technology (in press)

[9] Corrigan M, Adams K, Cook A (2007) Development of an interface for integration of communication and robotic play. In: RESNA Annual Conference, Phoenix, AZ

[10] Eberhardt S, Osborne J, Rahman T (2000) Classroom evaluation of the arlyn arm robotic workstation. Assistive Technology 12(2):132–143

[11] Howell R, Hay K (1989) Software-based access and control of robotic manipulators for severely physically disabled students. Journal of Artificial Intelligence in Education 1(1):53–72

[12] Karlan G, Nof S, Widmer N, McEwen I, Nail B (1988) Preliminary clinical evaluation of a prototype interactive robotic device (IRD-1). In: Proceedings of the International Conference on Agents and Artificial Intelligence, Montreal, QC

[13] Kronreif G, Kornfeld M, Prazak B, Mina S, Frst M (2007) Robot assistance in playful environment user trials and results. In: Proceedings of the IEEE International Conference on Robotics and Automation, Rome, Italy

[14] Kwee H, Quaedackers J, van de Bool E, Theeuwen L, Speth L (1999) POCUS project: Adapting the control of the Manus manipulator for persons with cerebral palsy. In: Proceedings of the International Conference on Rehabilitation Robotics (ICORR), Palo Alto, CA

[15] Kwee H, Quaedackers J, van de Bool E, Theeuwen L, Speth L (2002) Adapting the control of the Manus manipulator for persons with cerebral palsy: An exploratory study. Technology and Disability 14(1):31–42

[16] Musselwhite C, Wagner D, Cervantes O (2008) AAC authors: Writing beginning books for young readers. In: 13th Biennial Conference of International Society for Augmentative and Alternative Communication, Montreal, QC

[17] Nof S, Karlan G, Widmer N (1988) Development of a prototype interactive robotic device for use by multiply handicapped children. In: Proceedings of the International Conference on Agents and Artificial Intelligence, Montreal, QC

[18] Patrizia M, Claudio M, Leonardo G, Alessandro P (2009) A robotic toy for children with special needs: From requirement to design. In: Proceedings of the 11th International IEEE Conference on Rehabilitation Robotics, Kyoto, Japan, pp 918–923

[19] Poletz L, Encarnao P, Adams K, Cook A (2009) Robot skills of preschool children. In: RESNA Annual Conference, New Orleans, LA

[20] Robins B, Ferrari E, Dautenhaun K (2008) Developing scenarios for robot assisted play. In: Proceedings of the 17th Annual International Symposium on Robot and Human Interactive Communication, pp 180–186

[21] Smith J, Topping M (1996) The introduction of a robotic aid to drawing into a school for physically handicapped children: A case study. British Journal of Occupational Therapy 59(12):565–569

Chapter 5
Need- and Task-Based Design and Evaluation

Albert M. Cook, Jan Miller Polgar, and Nigel J. Livingston

Abstract Unfortunately, device abandonment (by clients or caregivers) is a pervasive problem in the provision of assistive technology. This is not necessarily the result of poor design of the technology, but rather a failure to account for other factors or determinants. This issue can be successfully addressed by employing the human activity assistive technology (HAAT) model when considering potential solutions for clients. The model conceptualizes the consumer, their activities, environment, and assistive technology as an integrated system in which changing one element affects all other elements in the system. The model can be applied in the design, selection, and evaluation of technology for use by an individual, or as a conceptual model for exploring the influence of assistive technology on participation in daily activities. In this chapter, examples and explanations are given for both "successful" and "failed" technologies with specific reference to the HAAT model.

5.1 Introduction

Well-intentioned but poorly designed assistive technology (AT) is a pervasive problem, and a contributing factor to device abandonment. By focusing on specific needs and specific tasks, AT design can be made much more efficient, and evaluation much more effective.

Albert M. Cook
Audiology and Speech Pathology, Faculty of Rehabilitation Medicine,
University of Alberta, AB, Canada, al.cook@ualberta.ca

Jan Miller Polgar
School of Occupational Therapy, University of Western Ontario, London, ON, Canada, jpolgar@uwo.ca

Nigel J. Livingston
CanAssist, University of Victoria, Victoria, BC, Canada, njl@uvic.ca

M.M.K. Oishi et al. (eds.), *Design and Use of Assistive Technology: Social, Technical, Ethical, and Economic Challenges*, DOI 10.1007/978-1-4419-7031-2_5, © Springer Science+Business Media, LLC 2010

5.2 Assistive Technology Abandonment

Technology abandonment occurs when a consumer stops using a device even though the need for which the device was obtained still exists. The underlying factors of device abandonment are important to consider. A survey of more than 200 users of assistive technologies identified four factors significantly related to the abandonment of assistive technologies: (1) failure of providers to take consumers' opinions into account, (2) easy device procurement, (3) poor device performance, and (4) changes in consumers' needs or priorities [4]. Another factor is the personal meaning attributed to assistive devices by the user [3].

Psychosocial and cultural variables can be primary factors in determining the meaning individuals assigned to AT. In particular, expectations of how the device would function and the social costs of using the device (i.e., cost/benefit of device use) determined whether a person integrated AT into his or her life or not. Personal characteristics related to mood, self-esteem, self-determination, motivation, and psychosocial characteristics related to friend and family support are also significant predictors of device use [5].

5.3 HAAT Model

The human activity assistive technology (HAAT) model conceptualizes the consumer, their activities, environment, and assistive technology as an integrated system in which changing one element affects all other elements in the system [2]. It can be applied in the design, selection and evaluation of technology for use by an individual, or as a conceptual model for exploring the influence of assistive technology on participation in daily activities.

The HAAT model has four elements: the activity, the human, the environment (context), and the assistive technology.

5.3.1 Human

The Human element considers all aspects that relate to the person, and is one of the most complex elements of the HAAT model. In considering potential assistive technologies, a consumer's cognitive, physical, and emotional skills and abilities as they relate to participation in daily activities and use of technology, are involved. Further, the consumer's experience with technology, whether they are an expert or a novice in its use, is also an important part of the human element.

5.3.2 Activity

Activity involves all those daily tasks in which a person wants or needs to engage. Often, these are conceptualized as self-care, productivity (work, education, volunteer activities), and leisure [6]. When considering the selection of assistive technology, it is important to determine which activities a person wants or needs to pursue, where these occur, and when they occur (e.g., time of day, frequency of occurrence).

5.3.3 Context

Context refers to different aspects of the environment that affect the person, the activities in which they engage, and their use of assistive technology. The conceptualization of the context involves four elements: physical, social, cultural, and institutional.

The physical environment includes both the natural and built environments as well as physical parameters (such as temperature and light) that affect the function and integrity of the assistive technology.

The social element of the context refers to other people and their interactions in the AT user's environment, and social conventions that affect technology use. The social environment can be conceptualized as a series of concentric rings representing increasingly distant relationships to the individual at the centre [1]. The first ring, the one closest to the individual, consists of family and close friends. The farthest ring consists of individuals or groups with whom the person has infrequent interaction, or whose actions affect the life of the person indirectly. The environments in which a person lives, works (or learns), plays and engages in community activities like shopping also form the social context. Here, the expectations of others and their attitudes and understanding of assistive technology affect whether it can be used successfully.

The cultural environment includes all those attitudes, beliefs, values, and practices that are shared by members of a broader group. Where a social group includes individuals who interact with each other with some frequency, a cultural group shares many values, but members may be geographically distant and unknown to each other. Cultural values such as independence and emphasis on physical abilities may discourage a person from using assistive technology that is seen as drawing attention to the user's lack of independence or physical skill.

Finally, the institutional element of the context refers to formal rules, procedures, policies, regulations, and legislation that affect daily life. Examples include the policies and procedures required to obtain funding for an assistive device or legislation related to accessibility of the physical environment and services to individuals with a wide range of disabilities.

5.3.4 Assistive Technology

The last element of the model is the assistive technology, specifically, the device and its interface with the user. Various features of the technology are considered in this element. These include the input required by the person when using the device, the interface between the device, the person and the environment, the device processor and the output of the device. Cook and Polgar [2] define hard and soft technologies, which refer to the actual device, in the first instance and the education, training, and other support needed to use the technology in the latter instance.

The design of an assistive technology system involves consideration of each of these elements, through a formal or informal evaluation. Cook and Polgar [2] describe a process to guide the design which involves starting at the activities that the individual wishes to perform when using the AT. The human skills and abilities, attitudes toward AT, and level of experience are considered next. The environments in which the activities are performed are identified as well as the need to transport the technology among environments. Once all of these elements and their interaction are considered, then selection of the actual device is considered. This process underscores the importance of assessing the needs and abilities of the person and the influence of the environment prior to determining which device is most useful.

5.4 Case Stories: Applying the HAAT Model

CanAssist is a program based at the University of Victoria (BC, Canada) that is dedicated to developing and providing technologies and services that increase the quality of life and independence of those with disabilities. In an approach consistent with the HAAT model framework, all projects undertaken by CanAssist are in direct response to requests – either directly from the individual with special needs or from a family member, caregiver, or health care-professional acting on their behalf. Before a project proceeds, the client coordinator talks with or preferably meets with the prospective user (and/or representative) to clearly understand the user's needs and the context of the request. The questions in Table 5.1: illustrate this data collection process.

The first question in Table 5.1: identifies the HAAT model activity to be performed. Questions 2 and 3 define the HAAT model context, as well as obtain information about the soft technology supports available. Question 4 begins to consider the assistive technology, especially the relationship between new and existing devices.

Once the information in Table 5.1: has been collected, a decision is made as to whether to proceed with the project. At this point, the critical questions focus on the potential assistive technology in the context of the HAAT model,

Table 5.1: Initial questions used to define a need

1.	What is the exact nature of the task that needs to be accomplished?
2.	Will the device be used in the home, school, or work setting?
3.	What supports will be in place for that individual?
4.	What other devices are in use, and will the new device in any way compromise the use of other devices?

and the existence of supporting soft technologies to ensure success. The primary determinant is whether the project is technically feasible. Additionally, CanAssist is focused on supplying devices when none exist commercially, so it is crucial to determine whether a solution already exists commercially. If a suitable device already exists, the client is directed to the source; if not, then a team will be assembled to develop the new device.

The client is a member of the design team, and is consulted on a continuous basis during the development of the device and after its delivery. To avoid the pitfalls associated with device abandonment, it is paramount that the client's expectations for performance and usefulness of the device are realistic and agreed upon by all parties.

5.4.1 Successful Projects

Keeping Marion connected at 92: Marion is 92 and has mild dementia. She lives in a care facility. CanAssist was approached by her daughter to develop an easy-to-use system by which she and her mother could regularly communicate (Marion is not able to use a regular phone). Thus, the activity is leisure and the activity output is communication. The context of use is primarily Marion's care facility.

In considering the development of the assistive technology component, the CanAssist team recognized that a broad constituency of users might have similar needs. The team's approach was to create a simple program based on Skype, a software application that lets people make free phone calls over the Internet. The new program, called CanConnect, simplifies the standard Skype user interface. CanConnect makes Skype accessible to people who have never used computers or are intimidated by them, and to people who are unable to use a regular mouse or keyboard. One benefit of using Skype is that it includes both audio and visual connections between the two parties.

To use CanConnect, Marion views a small gallery of photos, which are displayed on a computer screen. Using a touch-screen computer monitor, she simply touches the image of her daughter to establish audio and video contact with her. Other options for making audio and video contact include using an EMG headband, a webcam, or an eye tracking system. Marion uses CanConnect to have conversations with her daughter. Since its original development and customization for Marion, the CanConnect program has been distributed to many hundreds of users across Canada and internationally.

5.4.2 When Projects Don't Go So Well...

There are many factors that can lead to unsuccessful outcomes. Surprisingly, few of these are due to the technology not working as designed. In this section, we present several examples of how factors beyond the technology can affect success or failure of an approach. Often project failures arise from a lack of effective communication between the user (or the user's caregiver) and the design team, or from changed circumstances over the duration of the project. Sometimes the project can be rectified relatively easily (e.g., changing a mounting or the color of a device to make it more appealing or less obtrusive), but in some cases it is better to accept that mistakes have been made and that no further investment should be made in the project.

Sophie and the slippers: In some cases, needs are determined by focusing on the wrong person. Consider Sophie, who has Angelman Syndrome, characterized by profound developmental delay, a lack of speech, and ataxia of gait. Sophie's parents wanted a device that would encourage her to put on her slippers by herself with the expectation that once she did, she would also put on her shoes. This was considered to be a critical step in teaching Sophie to become more independent. Sophie loves music and dogs, so CanAssist modified a pair of slippers (featuring a large dog's head on the toes) so that when Sophie put both feet in the slippers her favorite song would play. (Each slipper is equipped with a pressure sensor, a radio receiver, and a transmitter, and a small MP3 player is embedded in one of the slippers.) Unfortunately, Sophie refused to put on the slippers and would insist that her parents activate her slippers themselves. The assistive technology worked flawlessly, the need was clear, and the reinforcement was matched to the client. Further, the way the reinforcement was built into the desired item was clever. So what went wrong? The critical element was that Sophie never participated directly in the design – rather, her parents decided that the slippers would be a good approach. The motivational component of the HAAT model's human element was not addressed. If Sophie had been brought into the process sooner, with a careful evaluation of why she did not put her slippers on, another approach might have been chosen.

Carol and the video camera: In some cases, a key person can wield control over the situation. Consider Carol, who has cerebral palsy. She was one of ten high school children with special needs who were each given an adapted video camera system. The camera system included control panels with large, easy to use buttons that control functions such as recording, zooming in and out, taking still photos, and turning the system on and off. Initially, Carol was extremely enthusiastic and produced wonderful images and footage with her system. Unfortunately, after a few months, the entire system was returned (and has since been given to another enthusiastic user). Under further examination, we found that Carol's original caregiver had moved on to a new position, and the replacement caregiver had a strong aversion to technology. The new caregiver was not able to provide Carol with the support and encouragement she needed to operate the system. Thus, when the soft technologies changed, the hard technologies could not be used. Carol's video camera illustrates the importance of having available the necessary soft technology supports.

5.5 Discussion and Conclusions

Obtaining a successful assistive technology outcome often hinges not on the technology itself, but rather on other factors incorporated into the HAAT model. The three case studies illustrate three factors that influence a successful outcome. Marion used the technology because it allowed her to continue with an activity (communication with her family) that was important to her. Sophie's slippers were not used because no one determined whether she wanted them, or the reason why she refused to use them. Carol was prevented from continuing with an activity she enjoyed because the environment no longer supported the use of her technology.

Soft technologies often determine success or failure of an assistive technology system. Focus on user needs and continued involvement of the user in the development process for custom assistive technologies is essential to avoid device abandonment. CanAssist maintains a very high success rate by paying attention to these critical factors, and by directly addressing the need for which the assistive technology should be designed.

References

[1] Blackstone S (2003) Social networks. Augmentative Communication News 15(1):1–16, http://www.augcominc.com/index.cfm/acn.htm
[2] Cook AM, Polgar JM (2008) Cook and Hussey's Assistive Technologies: Principles and Practice, 3rd edn. Elsevier, MO
[3] Louise-Bender Pape T, Kim J, Weiner B (2002) The shaping of individual meanings assigned to assistive technology: A review of personal factors. Disability and Rehabilitation 24(1–3):5–20
[4] Phillips B, Zhao H (1993) Predictors of assistive technology abandonment. Assistive Technology 5(1):36–45

[5] Scherer MJ, Sax C, Vanbiervliet A, Cushman L, Scherer J (2005) Predictors of assistive technology use: The importance of personal and psychosocial factors. Disability and Rehabilitation 27(21):1321–1331

[6] Townsend E, Stanton S, Law M, Polatajko M, Baptiste S, Thompson-Franson T, Kramer C, Swedlove F, Brintnell S, Campanile L (eds) (2002) Enabling Occupation: An Occupational Therapy Perspective. Canadian Association of Occupational Therapists, Ottawa, ON, http://www.caot.ca

Part II
Research and Academic Outreach

Chapter 6
Challenges to Effective Evaluation of Assistive Technology

Richard Simpson

Abstract Evaluation in the context of assistive technology can take several forms. Engineers evaluate the devices and technologies they develop. Clinicians perform clinical evaluations to decide which device(s) are most appropriate for their client. Finally, researchers hope to evaluate the long-term outcomes of assistive technology interventions. Evaluating a new assistive technology in the lab can be complicated by small user populations and the lack of universally accepted performance measures. The obstacles to effective clinical evaluations include the wide variety of devices that can meet each client's needs and the limited time and resources available to clinicians and their clients. Evaluating assistive technology outcomes is a relatively new pursuit, with many open questions.

6.1 Introduction

Evaluation is a critical aspect of the research and development process. The scientific method rests on the ability to measure the results of actions or interventions. Similarly, the development process is almost always depicted as a repeating cycle of implementation and evaluation. There are at least three different kinds of evaluation in regards to assistive technology (AT). Scientists and engineers evaluate the devices and technologies they develop as part of the research and development process. Clinicians perform clinical evaluations of a client's needs and abilities to decide which device(s) are most appropriate for their client. Finally, there is the idea (though rarely put into practice) of evaluating the long-term outcomes once a client has received AT.

Richard Simpson
Rehabilitation Sciences, University of Pittsburgh, PA, USA, ris20@pitt.edu

M.M.K. Oishi et al. (eds.), *Design and Use of Assistive Technology: Social, Technical, Ethical, and Economic Challenges*, DOI 10.1007/978-1-4419-7031-2_6, © Springer Science+Business Media, LLC 2010

6.2 Evaluating Technology in the Lab

Evaluating a new AT in the lab can be a challenge for even the most rigorous scientist with the best of intentions. Depending on the target user population, identifying a sufficient number of potential users who are able to travel to the lab to trial devices can be difficult. An alternative is to utilize single-subject or "small N" experimental designs, but these are difficult to generalize.

As an example, consider the DriveSafe System (DSS) [15]. The target user population for the DSS includes individuals who have both mobility impairments and visual impairments. This population is small to begin with and, almost by definition, has difficulty traveling to a lab. We were only able to successfully recruit two individuals from this population to test out the DSS. One approach we took to augment our testing with potential users was to use blindfolded able-bodied individuals as subjects.

Unfortunately, able-bodied subjects lack the orientation and mobility (O&M) skills of individuals who are actually visually impaired. A second approach was to blindfold actual O&M specialists, which was more realistic, but not even O&M specialists have the navigation skills of someone who is truly blind. A third approach we pursued was to use ambulatory blind individuals. These participants had the appropriate navigation skills, but they were unable to give us much insight into the needs of visually-impaired wheelchair users. In all three cases, we were able to recruit enough participants to perform group statistical analyses, at the expense of a realistic appraisal of the system's performance with the target user group.

A significant obstacle to accurately evaluating some technologies is the need for lots of training. Some technologies are inherently difficult to master, especially in cases where using a different technology outside of the lab in between training sessions can interfere with retention of skills taught during training. Even with regular use, some technologies can take months to master. For example, operating an augmentative and alternative communication (AAC) device with a sophisticated language encoding scheme like Semantic Compaction [1] is akin to learning a new language.

In the case of the DSS, some participants were given several hours of training prior to initiating data collection. Participants needed to learn how to operate a powered wheelchair that occasionally refused to travel in the indicated direction because of perceived obstacles. Then, participants needed to learn how to do this blindfolded.

A final obstacle to evaluating AT in the lab is the decision of what to measure. Investigators often emphasize speed at the expense of other valid measures like accuracy, comfort, or workload. When evaluating the DSS, for example, we knew the DSS was likely to cause participants to take longer to complete navigation tasks because the DSS slows down the wheelchair in the presence of obstacles. A participant completing a navigation task without the DSS could drive straight from the start point to the goal, if he or she chose, completing the navigation task in the shortest time possible at the expense of incurring multiple collisions. Our approach was to measure and report task completion time, collisions, and cognitive workload (measured through the NASA Task Load Index [4]) without trying to choose a single indicator upon which to base comparisons.

An alternative to evaluating AT in the lab is to evaluate AT "in the field" or "in the real world." Testing AT in unconstrained environments is appealing because that is how we ultimately hope it will be used, but there are challenges, as well. For example, there may be only one existing prototype of the device being tested, which makes field trials much less efficient than laboratory trials. In addition, real world environments are (by definition) less controlled than laboratories, making it difficult to replicate conditions across subjects. Field trials can also take much longer to generate data. For example, an entire day in a wheelchair user's life may feature much less driving than a single hour in a laboratory.

6.3 Evaluating Technology in the Clinic

Choosing the most appropriate AT is a collaborative decision-making process involving the consumer, clinician(s), and third-party payers. The challenges involved in a successful AT intervention include:

1. Evaluating and documenting client's goals, needs, and abilities [5].
2. Choosing the most appropriate AT to address these difficulties.
3. Configuring the technology to the user's needs.
4. Training the user in appropriate use of the system [10, 12, 14].
5. Providing continuous follow-up to ensure that the interface remains well-suited to the user [2, 9, 16, 14, 7].

The consequences of failing to successfully meet any one of these challenges include wasted human and material resources spent in the intervention process, unnecessary obstacles placed in the individual's path toward personal goals, and technology abandonment [9, 11].

Perhaps the greatest obstacle to effective clinical evaluation is the number of devices that are available. The fact that there are numerous mobility devices, communication devices, computer access devices, and other AT, each of which offers multiple configuration options, can be both a blessing and a curse. The existence of multiple products increases the odds that there exists at least one product that is well-suited for each consumer's specific needs. On the other hand, the variety of products and configuration options can be difficult to navigate when time (for both the clinician and the client) is constrained.

Clinicians are limited in the amount of time they can devote to a single client. Devices that can be trialled more efficiently may be emphasized over devices that require a lot of time for set-up in the clinic. Similarly, clients have a limited amount of time to devote to pursuing AT. Clients may have difficulty making repeated trips to a clinic to try out multiple devices and external deadlines, like school, training programs, or returning to work can intrude to force a decision. Consequently, a device that has an extended learning curve often loses out to a device that is easy to use quickly, even if the more complicated device might be a better choice in the long run.

A final obstacle to effectively evaluating a client's needs and abilities in the clinic is that both will change over time. A client's needs will change as job responsibilities, schoolwork, or personal interests evolve over time. A client's abilities can change due to the progression of their medical condition or changes in medication, but can also change as their familiarity with their AT increases.

6.4 Evaluating Technology in the World

Once a client has received an AT (or multiple AT), measuring the outcome of the entire AT intervention is extremely difficult. The discipline of AT Outcomes Measurement is still quite new and many fundamental questions remain unanswered, including:

- When do we measure? Immediately after the technology is received may be too early. The client may not have had adequate time to integrate the technology into his or her life. Do we wait until the technology has been in use for a while? Maybe, but how long? And, if we do, how do we isolate the AT intervention from everything else that has happened in the intervening time?
- What do we measure? Measures of performance can't be compared across device categories (e.g., mobility devices and communication devices). Even within a single device category, comparing performance measures across clients with different clinical goals (e.g., maximizing text entry rate vs. minimizing pain) may be fruitless. Measures of quality or satisfaction, on the other hand, can be difficult to define.
- How do we measure? Objective measures are attractive because they lend themselves to statistical analysis, but objective measures for inherently subjective concepts (e.g., quality of life) are not always readily available. Subjective measures, on the other hand, raise issues such as inter-rater reliability.
- Whom do we measure? The challenges presented by measuring AT outcomes are exacerbated by the diversity and (relative) scarcity of AT users. Identifying enough AT users to accommodate all of the potentially confounding variables, and isolating a control group of similar non-AT users, can be extremely difficult.

Several AT outcomes instruments have been developed, including the efficiency of assistive technology and services (EATS) [8], matching person and technology assessment (MPT) [13], psychosocial impact of assistive devices scale (PIADS) [6], and the Québec User Evaluation of Satisfaction with Assistive Technology (QUEST) [3]. However, none of these instruments measure the interaction of concurrent interventions or isolates the impact of an individual AT when multiple technologies are adopted at the same time.

6.5 Conclusions

There are challenges to evaluating assistive technology at all stages of product development and service delivery. Some of these obstacles are unavoidable when designing technology for a small, heterogeneous population with diverse needs. Others, however, may be overcome through advances in technique or technology.

References

[1] Baker B (1982) Minspeak. Byte 7:186–202
[2] Batavia AJ, Dillard D, Phillips B (1990) How to avoid technology abandonment. Washington, DC
[3] Demeres L, Weiss-Lambrou R, Ska B (1996) Development of the Québec user evaluation of satisfaction with assistive technology (QUEST). Assistive Technology 8(1):3–13
[4] Hart SG, Staveland LE (1988) Development of NASA-TLX (Task Load Index): Results of empirical and theoretical research. In: Hancock PA, Meshkati N (eds) Human Mental Workload, chap 7, pp 139–183. Elsevier, Amsterdam
[5] Hazell G, Colven D (2001) ACE centre telesupport for loan equipment. Oxford
[6] Jutai J, Day H (2002) Psychosocial impact of assistive devices scale (PIADS). Technology and Disability 14(3):107–111
[7] Lysley A, Colven D, Donegan M (1999) Catchnet final report. Oxford
[8] Persson J (1997) An overview of the EATS project: Effectiveness of assistive technology and services. In: Anogianakis G, Bühler C, Soede M (eds) Advancement of Assistive Technology, pp 48–52. IOS Press, Amsterdam
[9] Phillips B, Zhao H (1993) Predictors of assistive technology abandonment. Assistive Technology 5(1):36–45
[10] Raskind M (1993) Assistive technology and adults with learning disabilities: A blueprint for exploration and advancement. Learning Disability Quarterly 16(3):185–196

[11] Riemer-Reiss M, Wacker R (1999) Assistive technology use and abandonment among college students with disabilities. International Electronic
Journal for Leadership in Learning 3(23), http://www.ucalgary.ca/
iejll/riemer-reiss_wacker

[12] Scherer MJ (1993) What we know about women's technology use, avoidance, and abandonment. Women and Therapy 14(3–4):117–129

[13] Scherer MJ, Craddock G (2002) Matching person and technology (MPT)
assessment process. Technology and Disability 14(3):125–131

[14] Scherer MJ, Galvin JC (1996) An outcome perspective of quality
pathways to most appropriate technology. In: Scherer MJ, Galvin JC
(eds) Evaluating, Selecting and Using Appropriate Assistive Technology, chap 1, pp 1–26. Aspen Publishers, Gaithersburg, MD

[15] Sharma VK (2008) Design and evaluation of a distributed shared control,
navigation assistance system for power wheelchairs. PhD thesis, Department of Bioengineering, University of Pittsburgh, Pittsburgh, PA

[16] Tewey BP, Barnicle K, Perr A (1994) The wrong stuff. Mainstream
19(2):19–23

Chapter 7
Community Service in University Curricula

Nigel J. Livingston

Abstract CanAssist is a university-based program at the University of Victoria that is dedicated to developing and providing services and technologies to those with disabilities. All of CanAssist's activities are in response to requests from the community. The program has engaged many thousands of students (from a broad range of disciplines), faculty and staff, as well as many hundreds of community volunteers. The CanAssist model extends the traditional mandate of universities, to undertake research and education, by adding a third core pillar of activity – community service. This additional task is highly augmentative in that it creates outstanding new research avenues and provides students with a myriad of extraordinarily challenging and rewarding experiential learning opportunities.

7.1 Introduction

This paper provides an overview of CanAssist, a program dedicated to empowering those with disabilities, removing barriers to their full inclusion in society, and creating tools that will provide them with greater autonomy and independence.

A critical component of the CanAssist model is the extension of the traditional mandate of universities, to undertake research and education, by adding a third core pillar of activity – community service. This additional task, rather than reducing a university's capacity to fulfill its traditional role, is highly augmentative in that it creates outstanding new research avenues and provides students with a myriad of extraordinarily challenging and rewarding experiential learning opportunities. Beyond this, by directly

Nigel J. Livingston

CanAssist, University of Victoria, Victoria, BC, Canada, njl@uvic.ca

M.M.K. Oishi et al. (eds.), *Design and Use of Assistive Technology: Social,
Technical, Ethical, and Economic Challenges,* DOI 10.1007/978-1-4419-7031-2_7,
© Springer Science+Business Media, LLC 2010

addressing a key societal need, universities can further demonstrate their importance and relevance to society as a whole.

Another key element of the CanAssist model is the engagement of community. That is, CanAssist enables true partnerships between the university and the community at large, utilizing and leveraging the considerable combined resources of both. These partnerships encompass government at all levels – municipal, provincial and federal – and include business and industry, employers, charities, foundations and public sector agencies, as well as private individuals.

CanAssist is unique in being an interdisciplinary, university-based, service organization that develops and delivers technologies and services across the entire disability and age spectrum. To our knowledge, there are not any other similar organizations in North America.

7.2 The Need and Opportunity

Almost one in seven Canadians – roughly 4.4 million people – live with a disability. As Canada's population ages the proportion of people with disabilities will likely rise dramatically. Many of these Canadians face enormous challenges and barriers that not only make it difficult for them to accomplish the basic tasks of daily living but also severely limit their ability to participate in society.

Many of those with disabilities lack personalized equipment and technology that might allow them to better address the physical and/or cognitive challenges or limitations presented by their condition. Further, people with disabilities live in a world where too often little or no consideration is given to their special needs [3]. For example, consider the growing trend of miniaturization of devices (i.e., cell phones, digital and video cameras, MP3 players, etc.), while those devices are simultaneously becoming more feature laden. This combination makes operation of increasingly tiny and complex devices difficult for those who are lacking in dexterity, sight impaired, or facing cognitive challenges. Similarly, computer interfaces designed only for use with a regular keyboard and mouse (with attendant pull-down menus) as primary input devices are often not accessible to many people with disabilities. Thus, rather than benefiting from the wonderful opportunities offered by the Internet, many people with special needs are marginalized and excluded.

The fundamental issue behind inaccessible technology is the lack of economic incentives for the majority of manufacturers (e.g., of consumer electronic products) to address the needs of those with disabilities. Manufacturers often assume that the market is too small to warrant attention. Manufacturers who target, for example, a demographic of young successful professionals, may consider the production of fully accessible devices to be inconsistent with their brand or the image they wish to project [2]. Even those manufacturers who specialize in assistive technologies devote their attention to devices that have relatively large markets (wheelchairs, lifts, walkers, etc.) with the result that there is a paucity of companies that design technologies or devices that can be tailored to individual needs, and address some of the most basic needs of disabled people.

Apart from these "physical" barriers, a major impediment to disabled people being able to participate in society comes from a general lack of knowledge or understanding of disability issues. All too frequently, the able-bodied underestimate the contributions to society that people with a disability can, and more importantly, want to make. This assumption does not necessarily stem from a lack of good will but from inaccurate or ill-informed perceptions.

Indeed, these pervasive misconceptions have been shown to extend to the workplace. Potential employers, for example, tend to focus on the perceived risks and necessary accommodations in hiring those with disabilities, and do not appreciate the benefits that might accrue from having such motivated individuals working for them.

There is no doubt that perceptions can change through direct experience. Therefore, any program that can provide opportunities for extensive interaction between those with disabilities and those who will take a leadership role in shaping society has the potential to be truly transformative.

7.3 The CanAssist Model and History

CanAssist was founded just over 10 years ago at the University of Victoria in BC, Canada (originally, as the University of Victoria Assistive Technology Team (UVATT)). The original concept was to harness resources both within the university and in the greater community to address the unmet technological needs of those with disabilities. CanAssist was conceptualized as an organization that, given the extensive expertise and facilities available, would develop highly customized devices on an individualized basis to improve the quality of life of users. From the onset, a key imperative was that CanAssist would be a service-based organization, in that projects would only be undertaken in response to requests from the community. However, in delivering its services, CanAssist would utilize both educational and research resources and methods. This still holds true.

CanAssist started with a single project in 1999. For the first 5 years, all of its activities were undertaken on a volunteer basis by university staff, faculty, and members of the community. However, in order to provide reliable service and preserve continuity, permanent staff were eventually hired. Through donations from corporations and other philanthropists, CanAssist was able to hire seven full-time staff in 2005. Today, CanAssist has over 30 staff members (excluding faculty), including 17 engineers and computer scientists. In 2007, the university Senate and Board of Governors gave formal recognition to

CanAssist and appointed a full-time director. In addition, CanAssist has been given dedicated space (approximately 6,000 square feet) that includes laboratories, machine and fabrication shops and testing rooms.

CanAssist now receives approximately 200 requests a year for assistance. The majority come from individuals and families, although a significant number come from professional care givers and health care agencies. We develop over 40 new technologies a year, ranging from ball launchers (allowing disabled users to "throw" a ball for their dog) to electromyographic (EMG) based binary switches for software that allows users to easily browse the web without use of a mouse. Two particularly successful CanAssist technologies are the iPod adapter, PodWiz (Fig. 7.1), and an adjustable umbrella holder (Fig. 7.2).

Fig. 7.1 CanAssist's PodWiz iPod adapter has been provided to many hundreds of users with acute special needs. It allows them to control the music player with a wide variety of input devices including a single button switch (shown in the photo). The PodWiz also provides voice prompts, eliminating the need for users to read the iPod's display

One key ingredient to CanAssist's success is direct and regular contact with its clients. In fact, our philosophy is that our clients are valuable team members and their input is critical to the successful design of any device.

Another core tenet of the program is to engage students in CanAssist as widely as possible. One of the main anticipated benefits is that students, through their exposure to disability issues and the knowledge gained by having direct contact with disabled individuals, will be highly motivated to promote and exercise inclusion when in a position to do so. Of course, at the same time, students are provided with a unique, intellectually challenging and deeply rewarding experience that can only reinforce their feelings of good citizenship.

A third, and perhaps most critical element of the CanAssist model, is the engagement and partnership with community. This extends from the recruitment of individuals as volunteers to the forging of partnerships with community agencies, government, and industry.

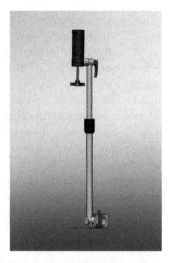

Fig. 7.2 CanAssist's adjustable umbrella holder is designed to be mounted on a walker or wheelchair to provide users protection from the sun or rain. Its design has gone through a number of iterations, primarily to reduce the cost of production and its weight

Through (a) the efforts of its regular full-time staff, (b) the recruitment of university faculty, staff, and students from virtually every discipline on campus, as well as community volunteers (typically, but not exclusively retirees), and (c) access to infrastructure support and facilities (e.g., machine shops, sophisticated laboratories, and equipment), CanAssist has become a significant resource for the disability community. It has been able to take on projects that are well beyond the scope of other agencies, either because of the technical challenges and/or the perceived lack of opportunity to recoup project costs.

While CanAssist is extraordinarily cost effective, program funding is a critical issue. We believe that through having multiple, diversified revenue sources (beyond infrastructural and direct support from the university) CanAssist can be sustainable. These sources include:

1. Service contracts with public and private sector agencies.
2. Funding from government (provincial and national).
3. Revenues accrued through the commercialization of technologies or delivery of other services.
4. Grants from traditional research funding agencies (e.g., provincial or federal agencies supporting scientific, medical, and engineering research).
5. Grants from philanthropic groups and organizations, corporate sponsors, and donations from individuals.
6. Social financing.

To date, CanAssist has not had a policy of charging individuals or families for its services. Rather, beneficiaries of the program who have some financial capacity are encouraged to make a donation should they see fit. In some cases, the cost of developing a specific customized technology can be very high. It might require, for example, the full time attention of two senior engineers for 6 months. It is not realistic to institute a charge directed at recouping a major portion of the cost of such a project. So within the program as a whole, there must be a means of subsidizing the more complex and challenging projects.

We also recognize, that over the long term, to be sustainable, we must secure revenues from some of our lower cost technologies, particularly those that might secure a relatively large market. There are a number of options to do so. One approach under consideration is to set up a small scale manufacturing facility that will employ people with disabilities in a number of facets of its operation (for example, in the processing of orders, production, packaging, inventory control). The benefits would be twofold: (1) accrued revenues would offset some of the costs associated with the development of individualized technologies and (2) meaningful and rewarding work would be provided to individuals, who by virtue of their challenges, have limited job opportunities.

Other opportunities to generate revenues will come about through the development of technologies that could be utilized in other arenas and markets (e.g., medical technology). This is particularly true of devices or products that could be adopted by general users, simply through the incorporation of good universal design features. For example, our CanConnect program, which allows calls to be made over the internet by simply touching an onscreen photo or icon, could appeal to a wide range of users who are not necessarily comfortable using a much more complicated computer interface [1].

7.4 The University Commitment: Engaging Faculty, Students, and the Community

In order for the CanAssist model to be adopted elsewhere, it is critical that universities make a fundamental commitment to the idea that community service should become an entrenched, recognized, and valued activity alongside research and education. Certainly, some work needs to be done to persuade administrators that more emphasis should be placed on community service when evaluating faculty, particularly with respect to determining tenure and promotion.

One key ingredient to the success of the CanAssist program is that even though its focus is on community service, it utilizes research and education to achieve its goals. Thus, there are numerous opportunities to involve faculty, whether through the supervision of students or through direct participation in projects and core research. Faculty involvement is not just limited to those in engineering, science or medical disciplines. For example, there are tremendous opportunities for those in the social sciences and in education to evaluate the benefits that accrue from the provision of assistive technologies from social or economic perspectives.

The CanAssist program provides a myriad of opportunities to engage students across all disciples, either through courses and practicums for which they can receive credit, or through the provision of volunteer projects. Thus, whole classes that undertake design courses (e.g., in mechanical engineering, computer science, and mechatronics) have been recruited to assist in the design of customized devices. In the process, students are introduced to the individuals who will receive the technology and their caregivers, thus giving a much better insight into the challenges faced by those with disabilities. Recently, commerce students and MBA students have been engaged in projects to investigate licensing and marketing of specific technologies.

Additionally, students are recruited for co-op work terms, honors projects and graduate studies. Students have the opportunity to work with individual clients on a one-on-one basis (mentoring the client and teaching them how to use a particular device) or as part of an interdisciplinary team.

We believe that local communities will continue to embrace opportunities to partner with the universities to address the needs of disabled people. Apart from the establishment of institutional partnerships and collaborations, a key element of community engagement is the recruitment of volunteers (many of these are retirees with a wide range of experience and background, including retired engineers, physicians, academics, machinists, seamstresses, and health care providers) to work in every facet of CanAssist activities, such as administrative tasks, research and development, and fundraising. We believe that with a well coordinated and professionally run program, volunteer contributions could account for up to 20% of a centre's activities.

7.5 Conclusion

CanAssist has engaged more than 3,000 students, over 200 faculty and staff, and approximately 200 community volunteers in its activities. In the process, it has developed over 150 novel technologies and provided direct assistance to many hundreds of clients. Thousands of others have benefited through the widespread distribution of software and computer programs. CanAssist has been able to flourish because the university has recognized that community service brings tremendous benefits not only to the community but also to the university itself, by creating outstanding experiential learning opportunities for students and also a myriad of research opportunities.

References

[1] CanAssist (2008) CanConnect. `http://www.canassist.ca/CanConnect`

[2] Goodman J, Langdon P, Clarkson P (2006) Providing strategic user information for designers: Methods and initial findings. In: Clarkson P, Langdon P, Robinson P (eds) Designing Accessible Technology, pp 41–51. Springer, London

[3] Lewis T, Langdon P, Clarkson P (2006) Investigating the role of experience in the use of consumer products. In: Clarkson P, Langdon P, Robinson P (eds) Designing Accessible Technology, pp 189–198. Springer, London

Chapter 8
Providing Innovative Engineering Solutions Between Academia and Industry

Brian E. Lewis and Yoky Matsuoka

Abstract Academic research is an invaluable development engine for creating assistive technology. Commercial enterprises are critical for making academically developed technology available to individuals provided that there are sufficient numbers of people with substantially similar technology needs. Unfortunately, there is little incentive for academic research organizations to optimize user interface, fit, finish, or function attributes associated with the assistive technology that they develop. The rate at which technology is being created that could be used to help people with disabilities is staggering. However, disturbingly little of that technology is being harnessed to provide actual benefit to the disabled community because the adaptation of that technology is viewed as too customized or the market size too small to justify commercialization. In this paper, the authors focus on the use of uniquely created non-profit organizations, such as YokyWorks Foundation, to bridge the gap between academic research and traditional commercial enterprises.

8.1 The Niche Between Academic and Commercial Approaches

There are many people with disabilities who could participate more fully in life, achieve more of their potential, and live more satisfying lives through technological assistance. Consider, for example, a college student with a C6 spinal cord injury who could not control his hands and fingers [5]. He had a brilliant mind, but his injury prevented him from writing, typing, grabbing

Brian E. Lewis
RosenLewis PLLC, Seattle, WA, USA, blewis@rosenlewis.com

Yoky Matsuoka
Computer Science, University of Washington, Seattle, WA, and YokyWorks Foundation, Kirkland, WA, yoky@cs.washington.edu

M.M.K. Oishi et al. (eds.), *Design and Use of Assistive Technology: Social, Technical, Ethical, and Economic Challenges*, DOI 10.1007/978-1-4419-7031-2_8, objects, and using his hands. He feared that effects of his injury would preclude meaningful employment. Moreover, he longed for the previously simple ability to eat food on his own without assistance. There was no medical or engineering intervention available on the market that could solve his problem.

Active academic research projects focus on understanding how human hands work [1] and how to augment these movements using technology [11, 4, 3]. However, development of such knowledge and technology is typically targeted toward solving research challenges and lacks design attributes that would make the technology practical for a daily user. In research, there is no incentive to build an easy-to-use device because investing time and resources reducing the weight or enhancing the design of a system does not address the fundamental research challenge of proving whether such a system is feasible, nor does it enhance the published output associated with that research. Sadly, for people suffering from these challenges, the technology developed through such academic efforts only proceeds through functional optimization if there is a market: a market is typically defined by the money that can be generated with developing a solution. In most cases, technologies for people with disabilities do not have a sufficient market because (1) the population is too small, (2) individual differences are too large, and (3) insurance does not cover devices that are life-enabling but not necessary for survival. YokyWorks Foundation [12] was created, staffed, funded, and organized to successfully solve problems such as these, that fall in the gap between academic research and commercial products.

8.2 History

YokyWorks grew from outreach projects undertaken at Carnegie Mellon University, personally funded and supported on a volunteer basis by students with a desire to have a positive impact on someone's life. The initial successes created referrals and inspired the team to search for existing organizations serving this niche. Looking for similarly situated organizations identified a void surrounded by three well established pillars: academia, commercial ventures, and philanthropic organizations.

Academic research institutions were perceived to focus on long horizon development efforts, and commercial enterprises on profitable and near term development efforts. Neither commercial nor academic enterprises are particularly well suited for near term research with highly uncertain commercial value. The challenge in developing individually oriented solutions using volunteers in an academic setting is multifold. Academic research projects often have long or uncertain development cycles, which is fundamentally inconsistent with the immediate needs that YokyWorks sought to address. YokyWorks sought to conduct much of the design, development and testing work using independent volunteers. Similarly, the commercial enterprise model does not work well for short term projects with a highly uncertain upside. Funding for commercial ventures typically requires either a short certain pathway to a return on investment or the potential for a large return on that investment. Because of the constraints inherent in both academic and commercial enterprises, the YokyWorks team needed to find an innovative approach.

Several non-profit organizations provide extraordinary benefit to people with disabilities. The Tetra Society of North America [9] has supported a large number of projects that help disabled people with specific challenges. Tetra typically focuses their efforts on direct adaptation of existing commercial devices tailored to the unique needs of a specific individual. Similar to Tetra is an organization in the United Kingdom called Remap [7] which creates unique pieces of special equipment tailor-made by volunteers and given at no cost to disabled individuals. In Australia, the Technical Aid to the Disabled of New South Wales (TADNSW) [8] also creates one of a kind solutions for disabled people in that region. In Canada, CanAssist [10], a university affiliated service organization, creates customized solutions for people with disabilities, and is described in Chap. 7 of this book. The Neil Squire Society [6], a nonprofit organization in Canada, offers personal consultation, adaptation, and worksite modification to existing technologies as part of its mission to facilitate social and economic independence for people with disabilities. YokyWorks saw the opportunity to tackle more challenging development closer to the academic research space with the potential to help more than the single individual initially being served.

YokyWorks' challenge was to blend some of the best attributes of academic and commercial structures with the benefits of non-profit structures. Historically, non-profit organizations have been used to provide services to people in need, but it is far less common for non-profit organizations to develop goods that benefit many people in need. The YokyWorks model does not work as a commercial enterprise because no sincere assurance could be made to an investor that there would ever be a return on investment. Compounding this was YokyWorks' objective to tap into the good will of volunteers to provide much of the brainpower behind the organization. Similarly, the YokyWorks model does not work in an academic setting because results needed to be achieved near term and were directed to people with immediate needs. Consequently, YokyWorks sought to use a non-profit model that incorporated certain elements of both commercial and academic enterprises.

8.3 Project Criteria

YokyWorks faces a problem common to both commercial and non-profit enterprises: It has more need than resources. The organization lacks the resources to help every person who requests assistance. The challenge for the organization is creating a repeatable analytic paradigm that efficiently sifts through the volume of requests to produce the most fruitful projects. YokyWorks seeks projects that meet two objectives: (1) there is no existing commercial solution and (2) providing a solution for this person might also provide a solution for other people. These objectives are explored in more detail below.

8.3.1 Absence of Existing Solutions

While YokyWorks is fundamentally in the business of engineering solutions to problems where no commercial solution exists, the organization has been surprised at the number of applicants for whom a solution is on the market. The most frequent reason YokyWorks turns projects away is that a solution already exists. Other non-profit organizations assist clients in finding appropriate existing assistive technology. Unfortunately, prospective clients are often unaware of these organizations and lack the familiarity to be able to search for these devices on their own. Accordingly, the first step for every project is to find a team member to look at the state of the art used to solve challenges similar to the one being evaluated. This effort amounts to looking for a null set in a diversely populated universe, and, as such, lacks complete certainty. Despite the lack of certainty, after a reasonable effort, it is possible for YokyWorks to conclude that no existing solution adequately addresses the challenge faced and that the case merits further evaluation.

8.3.2 Potential for Direct and Ancillary Benefit

YokyWorks is in an unusual niche. The organizational goal is to find projects that both benefit a specific individual and hold the potential to assist many additional people with a similar challenge. Simply, the organization wants to provide the greatest benefit from each dollar of funds it spends on research and development. Finding projects that can benefit a large number of people may seem incongruous with developing unique engineering solutions. If there are an enormous number of people with the same need, an existing commercial solution becomes a virtual certainty. Despite the apparent conflict between finding problems for which no solution exists and finding problems that could benefit many people, the organization has found several promising candidates.

8.4 Example Projects

8.4.1 Umbrella

One of the projects that led to the formation of YokyWorks involved a boy in a wheelchair with a fairly simple problem. It often rained where he went to school, and to cross campus involved using an umbrella. Commercially available umbrella holders for wheelchairs held the umbrella directly above the individual. This solution initially seems reasonable. Unfortunately, the student often found himself soaked and uncomfortable by the time he arrived at class. The gestalt moment in the project came with the fairly obvious realization that while we perceive rain to fall "down," the angle of incidence between rain and an individual after factoring in wind and velocity of travel is almost never straight down. The team developed a readily adjustable system that allowed the umbrella to be positioned on a wide variety of angles with different radial orientations. This enabled the student to quickly adapt the umbrella to the prevailing wind direction and velocity relative to his own direction and rate of travel. In the end, he spent far fewer days focusing on his physical discomfort in class, and more time learning.

8.4.2 Orthotic Exoskeleton

The opening remarks of this paper described the motivation for developing an exoskeletal pneumatically actuated assistive grip device. What the person suffering from vertebral injury needed was a simple power enhancing device for his hand without the level of sophistication of typical research exoskeletal devices. Our team designed a simple device that enables index/thumb pinching motion with a biceps EMG control (Fig. 8.1). The student was then able to grab different objects with his own volition without relying on the assistant, and could eat some soft food that he longed to eat on his own. We are proud that this project has both benefitted the individual, and lead to academic publications. Two undergraduate students who worked on this project have successfully written a conference and a journal paper [5, 2]. While this project was more research targeted than a typical YokyWorks project, it is a great example of the level of sophistication in engineering that YokyWorks focuses on.

YokyWorks is actively pursuing development of several innovative projects including a communication device for children with cerebral palsy and other movement disorders, and devices that make writing and eating easier for people with Parkinson's Disease.

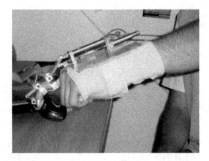

Fig. 8.1 Orthotic exoskeleton system devised to assist a person with spinal cord injury

8.5 Logistics

To build a successful foundation, YokyWorks has developed mechanisms for identifying and recruiting volunteers, managing teams of disparately located volunteers, funding development efforts, managing liability issues, and assuring intellectual property rights to sustain the organization.

8.5.1 Identifying and Recruiting Volunteers

One of the pleasant surprises while putting together YokyWorks was that a large perceived potential hurdle was easier to solve than expected. This is not to say that finding volunteers, identifying the particular skill sets required and keeping those individuals engaged over the course of a project is not challenging. It was anticipated that finding qualified people to perform testing, evaluation, engineering, and project management work on a volunteer basis could prove an insurmountable obstacle – these were busy professionals that had every reason to expect compensation for their services. The challenge was to identify skilled people interested in philanthropically providing service to people with disabilities through application of their unique skills.

In addition to finding a pool of skilled volunteers, there is the challenge of assigning from this pool people with the appropriate skill set to work on a particular project. It is clear that the skill sets of the people working on a problem will have an enormous impact on the nature of the solutions pursued. A physical therapist may be more inclined to pursue rehabilitative solutions while a mechanical engineer may look for augmentative solutions. Thus, developing a project team requires some early stage brainstorming of probable solutions, conceiving the largest technical roadblocks to achieving those solutions, and identifying the skill sets needed to address those challenges. Because each new project requires a unique combination of skills,

YokyWorks is constantly looking for people that can bring their skills to bear on potential challenges, and keeping those people engaged until a project that matches those skills arrives.

8.5.2 Funding

YokyWorks uses four approaches to funding its work: Seeking grants, accepting donations, managing expenses, and (at some point in the future) developing passive revenue sources. The principal near term sources of funding for the organization include a combination of grants and donations. The organization has consciously avoided the traditional funding source of commercial investors to avoid the concomitant requirement that the organization provide those investors a financial return on their investment. Grant funding is used where feasible and is expected to be a long term funding source for the organization. Private donor funding has been critical to launching the organization. Organizing the entity as a non-profit assists in the fundraising challenge by providing significant tax incentives to donors.

YokyWorks' volunteers make an invaluable contribution to meeting the funding needs of the organization. These volunteers contribute to funding by attacking and reducing the expense side of the business. It is perhaps obvious, but still worth stating: The lower the organization's expenses, the less time and energy must be invested in raising money for the organization and the greater the number of projects the organization can undertake.

It is conceivable that at some time in the future a third funding source may arise if the devices YokyWorks develops benefit enough people. People who benefit from YokyWorks devices may be willing to financially support the organization either directly or indirectly. That support may take the form of purchasing an industrially hardened embodiment of a YokyWorks device or making a direct donation to YokyWorks.

8.5.3 Liability Management

YokyWorks creates devices that people use in the real world under harsh and unpredictable conditions. The solutions we develop and test are often prototypes that have not been industrially hardened. These devices are faced with situations and factors that our team of volunteers could not possibly have imagined during design, and we design for some pretty amazing environmental conditions. For example, one YokyWorks project is developing a device to assist an individual whose ability to walk is severely compromised because he contracted poliomyelitis (polio) as a youth. Using the device, this person will seek to race across 151 miles of the Sahara Desert over 6 days in the Marathon Des Sables under his own power. YokyWorks' devices may be designed to give a person a wider range of motion, travel, and speed. The user may not be practiced in controlling such motion. In short, liability may arise from any number of sources.

Unfortunately, lawsuits are prevalent in the United States, so it is worth undertaking cost effective prophylactic strategies to minimize the risk and impact of potential litigation. Most of these practices are common in the industry. We use a non-profit corporation to shield individual volunteers and others supporting our work from personal liability, we use individualized statements of risk for each client tailored to the reasonably foreseeable risks, and we use volunteer agreements to make sure our relationship with volunteers is clear and well understood.

8.5.4 Intellectual Property Issues

YokyWorks is focused on solving technical problems for which no solution exists. If a project matures to produce a solution, that solution addresses one individual's problem. If the project is well selected, that solution may benefit other similarly situated individuals. Inherent in the YokyWorks model is the notion that the projects will produce new know-how and inventions. Because the people contributing to these inventions may come from universities or corporations there are commonly restrictions on creating patents that are not owned by the university or corporations.

Unfortunately, simply being a non-profit does not shield YokyWorks from such claims. Fortunately, many such claims are legally prohibited because states like Washington and California prevent claims by employers against work that is not related directly to the employer's business at the time of the invention, or inventions that cannot be shown to demonstrably result from work performed by the volunteer for the employer.

YokyWorks is still a young organization and it is unclear how it will leverage its intellectual property to effect the greatest good for the largest number of beneficiaries. Alternatives include putting specific devices into production or licensing the technology to people expert in manufacturing and marketing these services.

8.6 Conclusion

Blending the research and grant funding skills of academia, the team management and building skills common to for-profit enterprises, and the donors and volunteers that support non-profit companies has enabled YokyWorks to rapidly develop unique engineering solutions for people facing challenges in life. The organization is eager to grow and expand and take on as many new engineering challenges as possible.

Acknowledgements The authors gratefully acknowledge the contribution of our volunteers, donors, and clients who have trusted us to confront the challenges they face.

References

[1] Bradberry T, Gentili R, Contreras-Vidal (2010) Reconstructing three-dimensional hand movements from noninvasive electroencephalographic signals. Journal of Neuroscience 30(9):3432–3437, doi:10.1523/jneurosci.6107-09.2010

[2] DiCicco M, Lucas L, Matsuoka Y (2004) Comparison of control strategies for an EMG controlled orthotic exoskeleton for the hand. In: Proceedings of the IEEE International Conference on Robotics and Automation (ICRA), New Orleans, LA, pp 1622–1627

[3] Ehrsson HH, Rosen B, Stockselius S, Ragn C, Khler P, Lundborg G (2008) Upper limb amputees can be induced to experience a rubber hand as their own. Brain 131(12):3343–3352, doi:10.1093/brain/awn297

[4] Kipke DR, Shain W, Buzsaki G, Fetz E, Henderson J, Hetke J, Schalk G (2008) Advanced neurotechnologies for chronic neural interfaces: New horizons and clinical opportunities. Journal of Neuroscience 28(46):11830–11838

[5] Lucas L, DiCicco M, Matsuoka Y (2004) An EMG-controlled hand exoskeleton for natural pinching. Journal of Robotics and Mechatronics 16(5):482–488

[6] Neil Squire Society (2010) We use technology, knoweldge, and passion to empower Canadians with physical disabilities. http://www.neilsquire.ca

[7] Remap (2009) Custom made equipment for people with disabilities. http://www.remap.org.uk

[8] Technical Aid to the Disabled (2009) New South Wales. http://www.technicalaidnsw.org.au

[9] Tetra Society of North America (2009) http://www.tetrasociety.org

[10] University of Victoria (2008) CanAssist. http://www.canassist.ca

[11] Velliste M, Perel S, Spalding M, Whitford A, Schwartz A (1998) Cortical control of a prosthetic arm for self-feeding. Nature 453:1098–1101, doi:10.1038/nature06996

[12] YokyWorks Foundation (2010) Engineering for the human experience. http://www.yokyworks.org

Chapter 9
Case Study: An Assistive Technology Ethics Survey

Peter A. Danielson, Holly Longstaff, Rana Ahmad, H.F. Machiel Van der Loos, Ian M. Mitchell, and Meeko M.K. Oishi

Abstract This chapter describes the online *N-Reasons Ethics and Assistive Technology* survey (AT) designed to address key ethical issues in assistive technologies. The survey was used to foster deliberation and focus discussions in a multidisciplinary workshop on assistive technologies. The survey focused on each of the four workshop topics (evaluation, sensing, networking, and mobility). This chapter thus begins with an overview of the survey design in Sect. 9.1 followed by the process that was used to establish survey content in Sect. 9.2. The results for the survey are presented in Sect. 9.3 followed by brief conclusions in Sect. 9.4.

A survey on the ethics of assistive technologies was commissioned to identify debatable issues that could facilitate discussion about assistive technologies in a multidisciplinary setting. The survey was designed by a multidisciplinary group of researchers in assistive technology prior to the workshop. All workshop participants completed the survey, as well as members of the general public. The results of the survey provided data about which issues were non-controversial, and which issues were far less clear.

Peter A. Danielson, Holly Longstaff, and Rana Ahmad
Applied Ethics, University of British Columbia, Vancouver, BC, Canada,
`pad@ethics.ubc.ca`, `longstaff@interchange.ubc.ca`, `rahmad@interchange.ubc.ca`

H.F. Machiel Van der Loos
Dept. Mechanical Engineering, University of British Columbia, 6250 Applied Science Lane, Vancouver, BC, V6T 1Z4, Canada, `vdl@mech.ubc.ca`

Ian M. Mitchell
Dept. Computer Science, University of British Columbia, 201-2366 Main Mall, Vancouver, BC, V6T 1Z4, Canada, `mitchell@cs.ubc.ca`

Meeko M.K. Oishi
Electrical and Computer Engineering, University of British Columbia, 2332 Main Mall, Vancouver, BC, V6T 1Z4, Canada, `moishi@ece.ubc.ca`

M.M.K. Oishi et al. (eds.), *Design and Use of Assistive Technology: Social, Technical, Ethical, and Economic Challenges*, DOI 10.1007/978-1-4419-7031-2_9,
© Springer Science+Business Media, LLC 2010

9.1 Survey Design

The Ethics and Assistive Technology survey (AT) addressed key ethical issues in assistive technologies and employed the N-Reasons experimental online survey platform developed by the Norms Evolving in Response to Dilemmas (NERD) research team led by Peter Danielson at the University of British Columbia's Centre for Applied Ethics. This novel platform provides a means of engaging both the general public and experts in various ethically challenging issues and debates in two formats: (1) reason-based responses (described in greater detail below) and (2) the more conventional survey question formats (e.g., multiple choice, ranking) [1, 8, 2, 9, 6]. To date, the NERD research group has launched N-Reasons surveys on a wide variety of topics including research ethics, stem cell research, and robot ethics [9, 7, 5, 4].

The AT survey consists of five scenarios accompanied by one or more questions related to the various issues that each scenario involves. A total of 14 questions are posed, each with the option to answer "Yes," "Neutral" or "No." Participants must select one of these responses and provide a reason, explanation or elaboration to move forward through the survey. The innovative feature of the N-Reasons platform is the opportunity participants have to vote for other participants' reasons instead of (or in addition to) providing their own (see Fig. 9.1). The goal is to generate richer and more varied alternatives based on user-supplied contributions. The number of reasons the user chooses from (e.g., the "choice problem") is kept to a reasonable number by limiting content in three ways. First, by encouraging participants to use existing reasons rather than generating their own, the number of overall reasons is minimized and therefore more likely to result in identifiable trends or patterns. Second, running vote tallies for each reason are provided, which allows participants to factor in the valuation of the available reasons by other participants (e.g., no sums for decisions are displayed in order to make the reasons, as opposed to the "Yes"/"Neutral"/"No" decision, salient.) The display ranking method used in the survey gives some weight to recent contributions in order to mitigate the primacy effect; this method is discussed in more detail in [4] and shown in Fig. 9.1 below, where the third reason from the top (with a vote of 1.0) is displayed above one with 2.0 votes. Finally, each participant can vote for multiple reasons so that there is no need to generate conjoint reason responses: "I agree with R#101 and R#111."

The NERD research group generally designs each new survey with a background empirical investigation. For this survey, we explored the effect of identifying reasons by either their author's pseudonym or merely by a generated number that represents the reason anonymously. The participants were divided into two groups with cohort 0 viewing only numbers (N = 45) and cohort 1 viewing pseudonyms (N = 50); see Fig. 9.1. All participants viewed the same reasons; only the author's identifier (appended to each reason, as shown in Fig. 9.1) was varied.

Home

2. Sensing related scenario and survey question: your views

Sensing Scenario: Barbara has decided that she needs to provide greater supervision for her frail mother who is showing early signs of dementia. Barbara's mother does not want to leave her home, but has had a series of incidents that leave Barbara questioning whether her mother might inadvertently do herself harm (e.g. leaving the stove on, leaving food unrefrigerated). Barbara has a few choices in how she plans to cope with this situation. One option is to outfit her mother's home with sensors, which might include cameras that provide a live video feed and infrared sensors that detect (in real time) whether or not a person has entered a room. If her mother falls, for example, Barbara would be notified via email on her Blackberry. Alternatively, Barbara could move her mother into a full-time care facility. The reputable facility she has in mind has recently begun to monitor its patients and staff for potential acts of violence or aggression. Hence all residents must agree to be monitored in all common areas in order to live in the facility.

Question 2E: Is it acceptable for the care facility to monitor residents in their bedrooms and bathrooms?

Please select the reasons closest to your own views on this question. You may also submit additional reasons. The most preferred are listed first.

☐ 109 (27.5/79) **Neutral** because It depends on the type of sensors being used (video cameras, infrared, others), the security measures put into place to assure privacy is not violated, and on how the data is being analyzed before presented for human consumption. (**910**)

☐ 108 (17.0/79) **Neutral** because it depends on what the patients (or their legal guardians) have agreed to. Some may feel the tradeoff in security is worth the loss of privacy, others not. Their autonomy to agree to terms that make sense to them ought to be respected insofar as they are capable of it, and if not their legal guardians should be the ones to make that call. (**972**)

☐ 11 (1.0/79) **Neutral** because A question of Safety vs Privacy. If the person's mobility and independence have been assessed and they are not deemed safe especially for the bathing task then some sort of an easy to use alarm/audio system could be used to allow safe privacy. (**1804**)

☐ 2 (2.0/79) **Neutral** because I do not feel that bedroom and bathroom monitoring should by any means be the "default". Suites could instead be provided levels of monitoring capabilities, which could be adjusted to match the needs of the residents. For those who are still mobile and mentally capable there could be simple touch pads to activate 2-way audio or 2-way visual. Those with mobility issues may be provided a remote for such a system (like the old "I've fallen" commercial); and lastly, those whose mental faculties have reached a point where monitoring may be necessary... would be a decision between the legal guardians of the resident and the facility. In all cases, audio contact should be attempted first, then video contact attempted if patient is unresponsive. Additionally, the frequency of video monitoring should be logged with timestamps, time/date/room that camera was on, and those should be made available to their legal guardian so they can assure their relative is not being spied on excessively. (**1578**)

Fig. 9.1 N-Reasons survey visual presentation (from Cohort 0, who views reason numbers, the number listed at the end of each reason). For each reason, the first number indicates display rank (equal to the number of other reasons this reason was preferred over, minus the number of other reasons preferred over it), the second number indicates weighted number of votes (fractions arise due to split votes), and the third number represents the total number of votes.

☐ -1 (3.0/79) **Yes** because Only is very special cases where there is a high risk of injury to a resident and only with consent of the resident or, what is the more likely scenario, the legal representative of the resident (**1438**)

☐ -2 (4.5/79) **No** because although there are situations in which some form of monitoring may be necessary for the health and safety of the patient, these should be viewed as exceptions, required by extreme conditions with careful safeguards for approval in hard cases (e.g., where the patient is not capable of responding and there are no relatives or legal representatives with authority to sign or speak for the patient). (**995**)

☐ -17 (10.0/79) **No** because In an institution, there are other ways to monitor besides video. THis is extremely invasive to be monitored in the toilet. In a care home, staff are still critical for proper care at this level. Routine surveillence should be done by staff. (**886**)

☐ -28 (0.5/79) **Neutral** because every case is different there may be ceeeertaian situations were this might be nesssary but bathrooms and bedrooms are to invasive (**1032**)

☐ -46 (1.5/79) **No** because In general, No. However the final decision rests on mental health of the patient. If the patient is capable of making sound decisions, patient's wish takes precedent. If the patient is incapable of relational judgment, although consent might be irrelevant yet less intrusive monitoring technology might be more adopted. (**982**)

☐ -48 (6.5/79) **Yes** because care home residents' accommodation normally comprises only these rooms and these would be high risk areas in terms of the abusive scenarios. However, this is highly intrusive. It would be imperative that design of the system be sensitive to privacy, for example not using visual images in bathrooms. (**898**)

☐ -52 (2.0/79) **No** because The word 'monitor' needs to be made clear. As worded, it would be a clear violation of a resident's right to privacy. A resident may need assistance (this will be spelled out in the care plan), but 'monitoring' (how?) 'just in case' is no more permissible here than it would be in the resident's own home. (**945**)

☐ -54 (8.5/79) **Yes** because many accidents happen in these situations. Again, permission should be given by the resident or their representative. (**862**)

Or add a new reason

◯ Yes because

◯ No because

◯ Neutral because

[Submit reason(s)]

Fig. 9.1 (continued) N-Reasons survey visual presentation (from Cohort 0, who views reason numbers, the number listed at the end of each reason). For each reason, the first number indicates display rank (equal to the number of other reasons this reason was preferred over, minus the number of other reasons preferred over it), the second number indicates weighted number of votes (fractions arise due to split votes), and the third number represents the total number of votes

9.2 Survey Questions

Four topics were selected by a multidisciplinary working group in assistive technology at UBC prior to the workshop. They represent a set of topics considered not only highly relevant to assistive technology, but also intractable without multidisciplinary collaboration. The survey questions were designed to address each of the four workshop topics:

- **Evaluation:** How and why are assistive technologies being used, and what sensor technologies could provide accurate data to assess usage?
- **Sensing:** What ethical and privacy concerns might be raised by the vast amounts of personal data that computer-controlled assistive technologies can easily collect, and how might technologies incorporate features to address those concerns?
- **Networking:** How do assistive technologies impact the sense of self, agency, sense of privacy, and/or quality of life of users and the people in their social circles (family, friends, caregivers, others)?
- **Safety and Mobility:** How, if at all, can technological innovations improve or mitigate some of the ethical concerns surrounding powered wheelchairs and their potential for harm to the wheelchair user as well as to others in the environment?

Participants of the workshop were then asked to propose key issues and solutions to these particularly difficult problems.

To generate the survey questions, the NERD research group solicited input from the expert participants (the Advisory Committee to the PWIAS-ICICS workshop): each expert was asked to contribute scenario-based questions that they felt ought to be asked of the general public and which, in their opinion, represented key issues. After some initial feedback, the Advisory Committee was presented with an additional opportunity to comment on or revise the scenarios and questions. These revisions were then compiled and edited by the NERD research team to produce the final set of questions, which comprised the AT Survey. In addition, a fifth scenario regarding athletic performance was added, given the timeliness of the workshop with the 2010 Vancouver Olympics.

The AT survey was formally launched to the general public 3 weeks prior to the workshop. The twin objectives of the survey were to identify (1) key ethical issues in assistive technologies and (2) the most significant topics in each of the workshop theme areas. The survey scenarios and questions are presented in Appendix 1.

9.3 Results

A total of 95 people participated in the survey, including both the general public as well as researchers involved with the workshop. Of the

95 participants, 76 completed all questions in the survey (N = 35 from cohort 0, N = 41 from cohort 1). Survey results can be found online [3]; results
for two of the five scenarios are reproduced in Appendix 2.

9.3.1 Aggregated Results

The survey produced clear qualitative outputs. While participants could
choose from an often rich menu of reasons to support their decisions, their
votes aggregate to a set of social decisions on the "Yes," "Neutral," and "No"
options. Figure 9.2 summarizes these results.

This level of aggregation allows us to characterize the answers to various
questions in different ways. First, in questions 2A, 2B, 2C, 3A, and 4B the
"Yes" answer is a clear majority choice. In contrast to these clear decisions,
while question 1C has a bare majority "Yes," nearly as many voted "No"
and no one voted "Neutral." We can characterize 1C as the most controversial question. In contrast, question 4C has a similar "Yes" vote but with
far fewer "No" votes. Questions 1A, 1B, 2D, 2E, and 4A all had a plurality of "Neutral" votes, and question 3B was almost evenly divided between
"Neutral" and "No" votes. Question 5 was also quite controversial. Both
Questions 1A and 5, two of the most most controversial questions, are presented in full in Appendix 2.

These rough characterizations based on aggregative votes should be qualified in two ways. First, they can be refined by considering the additional
information provided by the reasons participants voted for. For example,
some "Neutral" reasons protest the formulation of the question. Second, while
we can characterize these distributions of answers as social decisions in the
clearest cases (like the majority "Yes" and "No" cases noted) this is less
clear in the plurality cases. Announcing a decision rule in advance would
strengthen these characterizations and move our device from survey to social
decision procedure.

Evaluation Scenario: This scenario focused on obligations between a university and a student with a disability. Most survey participants stated that
they would require additional information to determine the appropriate level
of accommodation a university should provide to a student with a disability
(1A, 1B). While most survey participants agreed that an occupational therapist should not consider cost to the university in deciding what assistive
device the student needs, some survey participants disagreed (1C). However,
despite disagreement, the most popular "Yes" and "No" responses both held
in common that the occupational therapist's primary obligation is to the
student.

Sensing Scenario: Survey participants most clearly agreed that in group
homes, surveillance of residents always requires their approval (2A), and that

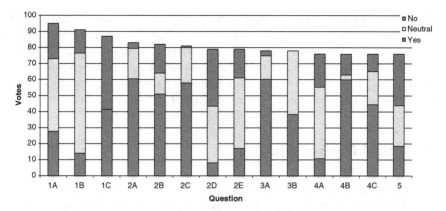

Fig. 9.2 Survey decisions by question. "No" is the *top dark shaded bar*, "Neutral" is the *middle lightly shaded bar*, and "Yes" is the *lower dark shaded bar*

in considering privacy of residents, raw data is more sensitive than data that has been read by machines and encoded into high-level characterizations (2B). Survey participants largely agreed that approval should also be required for in-home monitoring, as an integral part of respecting another person's autonomy (2C). The last two questions (regarding group homes) produced a variety of responses. For care facilities with residents who do not want to be monitored, the most popular response was that residents should not be removed, but rather accommodated as much as possible (2D). In considering whether monitoring of residents in bathrooms and bedrooms is acceptable, the most popular reason was "Neutral," depending on the type of sensors being used, on the degree of data encoding and manipulation prior to human analysis of the data, and on security measures put in place to prevent violations of privacy (2E).

Networking Scenario: Survey participants agreed that social inclusion should be a design factor (3A), along with other relevant design factors (e.g., cost, environmental impact, maintenance) (3B).

Mobility Scenario: This scenario focused on a powered wheelchair user living in a group home. Survey participants expressed a wide variety of opinions in assessing whether the occupational therapist should be able to reduce the wheelchair's maximum speed (4A); the most popular response was "Neutral" due to a lack of information. However, an unambiguous majority of survey participants believed that the user should be able to set the wheelchair's maximum speed (4B), with many responses citing the need for autonomy and personal choice of risk level. Many survey participants agreed that the group home should be able to set and enforce speed limits on their property; others pointed out the necessity of increasing wheelchair speed outside of the facility (4C).

Enhancement Scenario: While the majority of respondents believed that Oscar Pistorius should not be able to compete with able-bodied athletes, the reasons behind "Yes," "No," and "Neutral" responses varied widely (5). Reasons included the biomechanics of sprinting, comparison with other assistive devices, visibility of the prostheses, passivity of the device, Pistorius' skill, perceptions of fairness, and Pistorius' ability to inspire and attract viewers.

9.3.2 Self-Documentation

The survey is "self-documenting," meaning that both the quantitative and qualitative results are generated by the survey itself, thereby making the analysis both rapid and accurate. Immediate results can be obtained and updated as more users complete the survey. The results of both the Evaluation scenario and the Enhancement scenario can be found in Appendix 2 and the results from all 14 questions can be retrieved directly from the *yourviews* website [3].

9.4 Discussion

9.4.1 Survey Design

Each of the NERD surveys is an experiment in a broad sense, dependent on having enough voluntary participants. In a stronger sense, however, NERD also experiments on new methods in most surveys, aiming to improve our platform incrementally. In this survey, we added voting for multiple reasons as well as the modified popularity display ranking described above. We evaluate each of these innovations as a success. There were no complaints about the voting (as there had been in an earlier single vote survey) and our display method does mitigate the primacy effect as intended [4].

Finally, NERD conducted an experiment in the stronger sense of a random partition by dividing the population into two cohorts showing reasons identified by authors' pseudonyms or only by reason numbers. A preliminary analysis of these data suggests two small effects: first, those respondents who saw authors' pseudonyms contributed more reasons; second, they contributed more votes of "Yes" and less of "No," while neutral votes remained about the same. We are presently conducting further experiments on ways of linking authors and reasons, which we hope will help explain these differences.

9.4.2 Survey Content

Some questions produced clear agreement (2A, 2B, 3A, 4B), while others did not. One particularly useful element of the survey was the exploration of the respondents' reasons behind the simple "Yes," "No," "Neutral" vote. Agreement in assessment ("Yes," "No," "Neutral") was not necessarily synonymous with agreement in reasons. In some questions, participants agreed on the reason, yet came to different conclusions. In addition, in some questions with a high percentage of "Neutral" responses, many respondents stated that the question did not address the proper issue. This was still quite informative, since respondents often provided their own assessment of the relevant issue. Consider Question 1C, which falls into both of these categories:

> Suppose the University has in place an evaluation system in which a certified occupational therapist assesses the student's capabilities. Based on this assessment the needs of the student are determined and a recommendation is made to provide assistance. Should the occupational therapist consider the cost to the school when identifying assistance required for the student?

Question 1C elicited a variety of responses, including the following:

> **No** because the OT should make recommendations based on what is required of the student regardless of cost, but then someone OTHER than the OT should make a final decision that does take cost into consideration.
> **Yes** because the OT should make recommendations based on what is required of the student regardless of cost, but in high cost cases, also outline what can be achieved with a lower cost option and what limitations this places on the student. Then an informed decision can be made.

In this case, the reasons behind "No" and "Yes" revealed that the respondents believed that the question missed the relevant issue. While the question had been designed to elicit a prioritization of the student's or the school's needs, the issue of who ultimately made the informed decision was found to be more pertinent. Without the additional information provided in the reasons, "Yes" and "No" have far less meaning.

The results of the survey were incorporated into "seed" questions to facilitate roundtable, small-group, interdisciplinary discussions in (a) Evaluation, (b) Sensing, (c) Networking, and (d) Safety and Mobility. Some of the questions with high "Neutral" responses were useful in honing issues for discussion that would be of the most multidisciplinary interest. Some of the questions that did not have high "Neutral" responses, and were particularly divisive (5), were used as "icebreakers." Ultimately, workshop participants were given the leeway, as was found useful in the survey, to pose and answer questions they believed were most relevant. The initial list of questions identified for each discussion session is listed in Appendix 3.

9.5 Conclusion

The AT Survey was successful in generating both qualitative and quantitative results in response to the issues associated with assistive technologies and which formed much of the discussion during the workshop. Rather than merely producing one type of data or another, the survey has provided a more comprehensive set of data upon which further analysis can be performed.

It is possible to see from this approach that the issues involved are complex and include several factors to consider. In general, there were a number of neutral responses in all but two of the questions (1C and 4B), which are difficult to interpret in standard surveys; however, given the structure of the N-Reasons platform, it is possible to more clearly understand what the participants were concerned with and why they chose such a response. Additionally, the survey addressed those subjects that the experts engaged in this field felt were the most important and relevant to other researchers and the general public.

Acknowledgements This research was partially funded by the Peter Wall Institute for Advanced Research and Genome Canada through the offices of Genome British Columbia. Thanks to the NERD team, especially Robin Avery for programming the N-Reasons platform, and our participants for their enthusiastic support.

Appendix 1: Ethics and Assistive Technology Survey

- **Evaluation Scenario:** The University has purchased voice recognition software for students who have disabilities that make it difficult or painful to type. Jane is a student with this kind of disability and has been provided with the software but does not use it. She instead asks to be accommodated with a typist to whom she can dictate.

  1A. The software the University purchased was chosen for its accuracy and performance but also for its high degree of customization. Jane has tried to learn the software once or twice but she quickly gave up, finding it too difficult to learn. Do you think that the University has an obligation to accommodate Jane with a typist in this case?

  1B. There is a one-time cost associated with the software license which the University assumes will pay off over time. The cost of a typist is ongoing and dependent on a variety of factors such as the typist's availability and changing rate of pay. Do you think that Jane should be accommodated with a typist in this case?

  1C. Suppose the University has in place an evaluation system in which a certified occupational therapist assesses the student's capabilities. Based on this assessment the needs of the student are determined and a recommendation is made to provide assistance. Should the occupational therapist consider the cost to the school when identifying assistance required for the student?

- **Sensing Scenario:** Barbara has decided that she needs to provide greater supervision for her frail mother who is showing early signs of dementia. Barbara's mother does not want to leave her home, but has had a series of incidents that leave Barbara questioning whether her mother might inadvertently do herself harm (e.g., leaving the stove on, leaving food unrefrigerated). Barbara has a few choices in how she plans to cope with this situation. One option is to outfit her mother's home with sensors, which might include cameras that provide a live video feed and infrared sensors that detect (in real time) whether or not a person has entered a room. If her mother falls, for example, Barbara would be notified via email on her Blackberry. Alternatively, Barbara could move her mother into a full-time care facility. The reputable facility she has in mind has recently begun to monitor its patients and staff for potential acts of violence or aggression. Hence, all residents must agree to be monitored in all common areas in order to live in the facility.

2A. Should residential facilities be subject to ethical guidelines for this type of surveillance of their residents, to ensure the residents' approval, at some level, of the intrusion on their privacy?

2B. If personal data collected from assistive technologies is encoded and read only by machines instead of people, is it less sensitive? For example, real-time data (like video) would not be stored. Instead, only higher-level information would be extracted from it (e.g., a time-stamped event such as "subject went to living room").

2C. Should Barbara ask for her mother's permission to install sensors in her home?

2D. Does the care facility have the right to remove current residents (who moved into the facility before monitoring was implemented) if they are unwilling to be monitored?

2E. Is it acceptable for the care facility to monitor residents in their bedrooms and bathrooms?

- **Networking Scenario:** A university is built on the side of a steep hill. The university is deciding how to design access routes between buildings. One option is to build a series of long ramped paths, which can be used by everyone. The other is to build a staircase that goes up the middle of the hill for non-wheelchair users and a series of lifts which could only be operated with a key, and which could fit only a single wheelchair user and nobody else at one time.

3A. Should the social interactions of people using the potential access route be incorporated into the design of the university? In this case, should the design of the university aim to keep wheelchair users and non-wheelchair users on the same route?

3B. Should the university take into account the long term increased energy and maintenance costs associated with keeping wheelchair users and non-wheelchair users on the same route?

- **Safety and Mobility Scenario:** Peter is a 30-year-old intelligent man who has cerebral palsy with severe spasticity, which renders him unable to walk. He has limited fine motor control of one arm. In the past, he has tried a mouth switch and a head switch on a power wheelchair, as well as a standard joystick. However, therapists are reluctant to give him a power chair because he lives in a busy area of the city. Their fear is that he will either drive into a building or a person, or drive off the sidewalk and hurt himself. Currently, when he leaves the group home where he lives, he must be accompanied by an assistant who pushes him in a manual wheelchair. Peter has just been informed that a wheelchair manufacturing company is developing a prototype of a new power wheelchair that is maneuverable, accessible to a variety of user inputs (e.g., sip and puff, joystick), and has some safety features built in (e.g., bumpers to protect walls, furniture, etc. in soft collisions). Due to damage and liability concerns, the manufacturer is also planning to add a speed control function (accessible by a key code only) that determines the maximum possible speed.

4A. Peter's occupational therapist has recommended he try the new prototype wheelchair that includes safety features. The chair is outfitted with a "black-box" that continuously records maximum speed, average speed, and number of collisions. After a 30-day probationary period, the therapist evaluates Peter's driving record. Based on the black-box data, the therapist decides that Peter can continue to use the powered wheelchair, but that the wheelchair's maximum allowable speed will be set to half its previous value. Is this fair?

4B. Should Peter be given the key code to his own wheelchair?

4C. Assume that Peter will not have access to codes on his wheelchair. Does his group home have the right to dictate that only wheelchairs with speed control functionality can be used in the facility?

- **Enhancement Scenario:** "Despite having both lower legs amputated as a child, South African runner Oscar Pistorius dreamed of one day competing in the Olympic Games. That dream was dashed in early 2008 when the International Association of Athletics Federations ruled him ineligible, claiming his carbon-fibre prosthetics gave him an unfair advantage over able-bodied competitors. Pistorius appealed to the Court of Arbitration for Sport, which overturned the decision just in time for the Beijing Games. Unfortunately, the athlete known as Blade Runner fell seven-tenths of a second short of the Olympic qualifying time in the 400 m." [10]

5. Should Oscar Pistorius be allowed to compete with able-bodied athletes?

Appendix 2: Results of Question 1A and Question 5

Question 1A: The university has purchased voice recognition software for students who have disabilities that make it difficult or painful to type. Jane is a student with this kind of disability and has been provided with the software but does not use it. She instead asks to be accommodated with a typist to whom she can dictate (Fig. 9.3).

The software the university purchased was chosen for its accuracy and performance but also for its high degree of customization. Jane has tried to learn the software once or twice but she quickly gave up, finding it too difficult to learn. Do you think that the university has an obligation to accommodate Jane with a typist in this case? (Table 9.1)

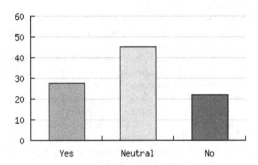

Fig. 9.3 Responses for Question 1A

Table 9.1: Participant reasons generated by Question 1A

Score	Reason
191 (28.5/95)	**Neutral** because not enough information is provided about Jane's difficulty with the software or the university's efforts to help her use it effectively. I don't want to waste time speculating on either, I prefer fuller information in the question. I don't think the university should accommodate someone who has not made an honest effort to participate, but there is no way of telling that.
42 (17.8/95)	**Yes** because Jane should be given a typist in the interim with encouragement to try modifying the software to improve success. In the long run she could be more independent if she found something that worked for her beyond a typist.
−3 (13.0/95)	**No** because more can be done to make the software usable for Jane. E.g., the university could offer a customization session. The typist approach will likely be very expensive long-term, and the most cost-effective option should be used (taking into account all costs, not just financial ones – Jane's frustration should be counted as a cost).
−9 (10.5/95)	**Neutral** because I am not sure how well the software performs. If it is well designed in terms of usability and demonstrates robust voice recognition then Jane should be strongly encouraged to persevere and not be given a typist unless she really has made a lot of effort.
−45 (7.0/95)	**Neutral** because Jane should be given a typist in the interim with encouragement to try modifying the software to improve success. In the long run she could be more independent if she found something that worked for her beyond a typist.

(continued)

Table 9.1: (continued)

Score	Reason
6 (6.7/95)	**No** because Jane has not made a good-faith effort to use the software. She will be more productive in the long term if she can do inputting into the computer without having to schedule a typist each time.
−69 (5.3/95)	**Yes** because this may be a difference between US and Canadian law. Under US law, the university must provide an effective accommodation for Jane, and as everyone who has used voice recognition software knows, it does not work well in some contexts, especially for someone who does not articulate uniformly. The reason given above does not state what more can be done, other than a customization session which will not be effective if Jane does not articulate uniformly.
8 (3.5/95)	**Yes** because accommodations cannot be "one size fits all." The technology the university is offering may not be suitable for Jane. If there is another viable or realistic alternative, such as a typist, it should be considered.
−9 (2.3/95)	**Neutral** because there is not enough information to tell if voice recognition software is appropriate for her, or if other technology would be more appropriate, a typist should be provided until a complete assessment is completed, voice recognition does not work for all individuals.
−69 (2.0/95)	**Yes** because not all people are comfortable using computers and the software might not be adaptable to all.
−12 (2.0/95)	**No** because first they should provide training. If this does not work, they should provide the typist.
25 (1.5/95)	**Neutral** because it is hard to tell the degree of disability from the scenario. Is Jane having a short term upper hand impairment? Is she in wheelchair with paralysis with decreased muscle strength in her hands? It is unclear whether the university conducted an assessment prior to giving Jane the equipment to check if it would be suitable for her. I am assuming that the university does not have a high number of students similar to Jane's needs, if indeed an assessment was not done, and Jane's disability is a chronic and debilitating one, the university ought to provide Jane with a typist while they fix the software issue.
−74 (1.0/95)	**Yes** because with the current state of technology, even the most advanced speech recognition software cannot adequately address Jane's needs. Contribution to software development is not Jane's responsibility.
15 (1.0/95)	**Yes** because I work with LD students and have an LD myself.
14 (0.8/95)	**No** because where do you draw the line in "assistance"? This could be an expensive precedent to set. Also, I personally have been unable to work due to a injury to my arm (I cannot type); my employers were not obligated to purchase me software to assist me and this is my livelihood, but I understand the cost factors involved if this were required.
−40 (0.5/95)	**No** because the software is designed for people with this disability.

Question 5: "Despite having both lower legs amputated as a child, South African runner Oscar Pistorius dreamed of one day competing in the Olympic Games. That dream was dashed in early 2008 when the International Association of Athletics Federations ruled him ineligible, claiming his carbon-fibre prosthetics gave him an unfair advantage over able-bodied competitors. Pistorius appealed to the Court of Arbitration for Sport, which overturned

the decision just in time for the Beijing Games (Fig. 9.4). Unfortunately, the athlete known as Blade Runner fell seven-tenths of a second short of the Olympic qualifying time in the 400 m." [10] (Table 9.2)

Should Oscar Pistorius be allowed to compete with able-bodied athletes?

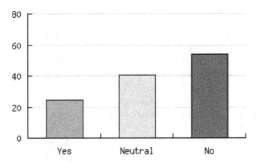

Fig. 9.4 Responses for Question 5

Table 9.2: Participant reasons generated by Question 5

Score	Reason
239 (23.0/76)	**No** because with prosthetics, the biomechanics of sprinting are significantly different than without prosthetics. A 400 m with prosthetics is a different sport than a 400 m without.
119 (13.0/76)	**No** because although I'm sympathetic to Pistorius' goals, there's no principled way to draw a line between Pistorius' blades and other assistive devices that would clearly give an unfair advantage.
113 (12.3/76)	**Neutral** because it is not clear whether Pistorius' prothesis is an "external device or piece of equipment" like a spring loaded shoe or whether it is an integral part of his body. Would an athlete with an artificial internal hip or knee joint be restricted from participating? Is it because the prosthesis is external and visible that we are considering discriminating against Pistorius? Wouldn't we want the most current technology in an internal knee joint for an athlete? Why not in an external prosthesis also? Or, do we view the prosthesis as we would a wheelchair which is a clear advantage for some running events over natural runners, i.e., it is an external device that is not part of the person? This is the reason given above for a Yes answer (acook), but its indeterminacy really supports a neutral stance.
−23 (9.0/76)	**Yes** because there is as yet no evidence that the prosthetics give him an unfair advantage. This is obviously a grey area – few would argue that he shouldn't be allowed to compete with no prosthetic at all, but most argue that he shouldn't be allowed to compete if his prosthetics were fuel-powered. A fully passive prosthetic is roughly equivalent to the introduction of the clap skate in speed skating – some skaters were faster with the clap skate than the conventional, but the top athletes were initially faster with the conventional skate. It would be very difficult to design a passive prosthetic which would confer any significant advantage over an intact athlete.

(continued)

Table 9.2: (continued)

Score	Reason
−26 (4.8/76)	**Neutral** because I do not know enough about the speeds achievable with the prosthetic limbs compared to able bodied athletes. I assume they do not infer an advantage and so should be allowed. However, if the prosthetic limbs, on average, increase performance then their user becomes a different class of athlete and should compete in a separate competition.
33 (4.0/76)	**Neutral** because in the same way that there are restrictions on swim suit designs, racket designs, there also has to be restrictions and rules for prosthetic design used in competition so it is not a greater advantage to have a prosthetic limb.
−26 (3.0/76)	**Yes** because we are speaking about a game. But then all able-bodied runners should be allowed to use carbon-fiber prosthetics, to be fair.
−53 (2.5/76)	**Yes** because agree with john. Also, perhaps Pistorius is a black swan compared to his peers. Now if all amputee runners began to post times better than olympics runners, then this would clearly be a different class. But the reality is that they are not even close and Pistorius just happens to be that good.
−73 (2.5/76)	**No** because even a passive prosthetic can provide an unfair advantage over other athletes. Where do we draw the line between Pistorius' blades and a bicycle fitted for amputated legs? What about spring-loaded shoes on able-bodied athletes? Should those be allowed? It seems that the best response at this point is to disallow all prosthetics which could potentially offer performance enhancements. One significant consequence of saying "no" here is questioning where the line of prosthetic enhancement ends (e.g., shoes).
−51 (2.0/76)	**Yes** because it is not clear whether Pistorius' prothesis is an "external device or piece of equipment" like a spring loaded shoe or whether it is an integral part of his body. Would an athlete with an artificial internal hip or knee joint be restricted from participating? Is it because the prosthesis is external and visible that we are considering discriminating agains Pistorius? Wouldn't we want the most current technology in an internal knee joint for an athlete? Why not in an external prosthesis also? Or, do we view the prosthesis as we would a wheelchair which is clear advantage for some running events over natural runners, i.e., it is an external device that is not part of the person?
−6 (2.0/76)	**Yes** because it's inspiring for other amputees and it would attract a large viewing audience. Fairness is all relative, and we actually make the rules so that the competition is interesting. Clearly, the blades do give an advantage to Oscar (I'm very familiar with this technology) and eventually prostheses or exoskeletons will enable Oscar and other athletes with physical "disabilities" to outperform their intact counterparts. As this happens, the Olympic committee will have to be creative in coming up with new rules that meet the spectators' expectations that the competition be both "fair" and all-inclusive. Perhaps separate categories could be created, each having their own technology-based rules. It may be that NASCAR could serve as a model for some future Olympic events.
9 (2.0/76)	**Neutral** because I do not know enough about the issue.
31 (1.5/76)	**Neutral** because if he can qualify with the devices he should be allowed.

(continued)

Table 9.2: (continued)

Score	Reason
−36 (1.3/76)	**Neutral** because the restriction should be based on how much the artificial parts enhance the performance of an athlete for a specific event. It's an inexact scientific assignment. It's clear that as science advances super mechanical parts will become available and no one argues athletes fitted with such parts should be allowed to compete.
−53 (1.0/76)	**No** because the right judges in such cases are those sponsoring the competition.
−43 (1.0/76)	**Yes** because the Court Of Arbitration For Sport is the supreme sports court. (CAS said the IAAF failed to prove that Pistorius' running blades give him an advantage.)
−36 (1.0/76)	**No** because his goal was to demonstrate that the accident did not affect his ability to run. He does not need to compete with professional athletes in order to do this.
−17 (0.5/76)	**No** because no maybe he should try olympics for the disabled.

Appendix 3: Small-Group Discussion Questions in the PWIAS-ICICS Workshop

- **Evaluation**

  1. What devices, methods, and protocols can assist researchers and clinicians in measuring how, when, and under what circumstances AT is being used? Should users make these determinations?
  2. What devices could be designed and implemented to bypass self-reporting? What best-practices might prevent violations of privacy if self-reporting is eliminated?
  3. What novel devices and methods for collecting data can be used to evaluate the impact of AT? What do we mean by impact?
  4. What novel devices, algorithms, and methods could non-intrusively detect/predict abandonment?
  5. How does device novelty affect its knowledge translation? Where in the pipeline from academic research to end-user use does knowledge translation fail, and why?
  6. When is AT appropriate? What user circumstances determine if AT should be used? When a device is abandoned, what determines whether a replacement device is required?

- **Sensing**

  1. How does a user's specific circumstance (type of disability, social network, use of AT) influence what type of data should be gathered?
  2. What level of security is required for user data? What privacy standards should be enforced? What are potential consequences of breaches?
  3. How much user benefit is required to overcome a loss in privacy (e.g., utility of Google mail often outweighs privacy concerns)?

4. How much control should users have over their sensor data? Why are specific types of sensor data more acceptable from a user's point of view than others?

5. Ubiquitous sensing technologies may require additional computing, storage, and communications infrastructure. How could this burden be mitigated to prevent potential derailment of new AT?

6. Ubiquitous sensing may be a deterrent for some people, but if designed well, could be desirable (e.g., an iPhone-based application for route finding via wheeled mobility). When is such technology desirable, or detracting, from the user's point of view?

- **Networking**

1. How does level of customization of AT affect a user's sense of agency? Are "generic" technologies less beneficial/useful?

2. How should concerns about agency impact technology design and use?

3. How can AT be designed to physically prevent or deter violations of a user's privacy or personal space?

4. Novelty and a steep learning curve can be significant barriers to the adoption of AT. How does switching to a new type of AT impact a user's sense of self?

5. How can novel technologies be designed to integrate seamlessly into the physical and social environments of the users and their surroundings?

6. To what end can sophisticated technologies be made user-friendly for people who are unfamiliar with computers, etc. (e.g., the elderly)?

7. What best-practices can ensure that devices are made from a need-based pull, as opposed to a technology-push?

- **Safety and Mobility**

1. What design practices can make AT more easily and reliably customizable to individual users?

2. What factors determine when mobility AT is warranted for a specific individual?

3. Under what circumstances should information recorded about the safety of a person's previous mobility behaviors be used to restrict or enhance the future capabilities of their mobility AT?

4. What best-practices in device and algorithm design could make mobility AT more robust to obsolescence? Given the high cost of mobility AT, what technologies would be required to make mobility AT modular and upgradeable?

5. What infrastructure should be developed to enable people who use mobility AT (e.g., Segways that may not be operated on sidewalks, powered wheelchair restrictions in group homes)?

6. How much control should group home residents have over their powered mobility AT? What other mechanisms could assure safe driving without sacrificing user autonomy?

References

[1] Ahmad R, Bailey J, Bornik Z, Danielson P, Dowlatabadi H, Levy E, Longstaff H (2006) A web-based instrument to model social norms: NERD design and results. Integrated Assessment 6(2):9–36

[2] Ahmad R, Bailey J, Danielson P (2008) Analysis of an innovative survey platform: comparison of the public's responses to human health and salmon genomics surveys. Public Understanding of Science Prepublished doi:10.1177/0963662508091806

[3] Centre for Applied Ethics (2010) Assistive technology survey results. http://www.yourviews.ubc.ca/en/AT_Survey_Results

[4] Danielson P (2009) N-Reasons: Computer mediated ethical decision support for public participation. In: Public and Emerging Technologies: Theorizing Participation, Banff, Canada

[5] Danielson P (2009) Survey on ethics and assistive technologies: N-reasons design and preliminary results. Presentation at the Peter Wall Institute for Advanced Studies Workshop, "Removing barriers and enabling individuals: Ethics, design, and use of assistive technology"

[6] Danielson P (2010) A collaborative platform for experiments in ethics and technology. In: van de Poel I, Goldberg D (eds) Philosophy and Engineering: An Emerging Agenda, Philosphy of Science, vol 2, Springer

[7] Danielson P (2010) Designing a machine for learning about the ethics of robotics: The N-Reasons platform. Ethics and Information Technology, Special Issue on Robot Ethics and Human Ethics DOI 10.1007/s10676-009-9214-x

[8] Danielson P, Ahmad R, Bornik Z, Dowlatabadi H, Levy E (2007) Deep, cheap, and improvable: Dynamic democratic norms and the ethics of biotechnology. In: Ethics and the Life Sciences, Journal of Philosphical Research, Charlottesville, VA, pp 315–326

[9] Ormandy E, Schuppli C, Weary D (2008) Changing patterns in the use of research animals versus public attitudes: Potential conflict. Poster at the 2008 International GE[3]LS Symposium, Vancouver, Canada

[10] Ripley S (2009) See you in court! Most memorable legal battles in sports. http://slam.canoe.ca/Slam/Top10/2009/05/24/9553531-sun.html

Part III
Development and Commercialization

Chapter 10
The Fast Pace of New Emerging Information and Communication Technologies: The Need for Regulations and Standards

Gary E. Birch

Abstract Information and communication technologies (ICT) are directly or indirectly a form of assistive technology. Indeed, many forms of new emerging ICT have the potential to provide opportunities for persons with disabilities to obtain inclusion in the mainstream of society to an extent never before accomplished. However, due to the rapid and ubiquitous insertion of these technologies into our society and given that most of these technologies are not accessible to persons with disabilities they are forming new barriers rather than opening up new opportunities. New strategies need to be adopted to deal with this trend and, particularly in the relatively near term, it is argued that the most important strategy is the development of enforceable regulations and standards that impact the accessibility of these new emerging technologies.

10.1 Introduction

Many emerging information and communication technologies (ICTs) have significant potential to enable persons with disabilities and promote their inclusion in mainstream society. This is in part due to the ever-increasing power of these devices to accomplish various technological tasks. The most significant factor, however, is that these new technologies are being incorporated into the mainstream of our society almost ubiquitously, including in areas such as employment, recreation, and various forms of service delivery. Combined with the fact that these technologies are changing at a very rapid pace, this lays the foundation for a perfect storm of inaccessibility for persons with disabilities. As new technologies come onto the market, they are not inherently accessible for a large majority of persons with disabilities and therefore become barriers rather than tools that allow them to take part in

Gary E. Birch
The Neil Squire Society, Burnaby, BC, Canada, garyb@neilsquire.ca

M.M.K. Oishi et al. (eds.), *Design and Use of Assistive Technology: Social, Technical, Ethical, and Economic Challenges*, DOI 10.1007/978-1-4419-7031-2_10,
© Springer Science+Business Media, LLC 2010

society like their able-bodied counterparts. The old paradigm of developing retrofit solutions to access mainstream ICT is failing persons with disabilities because in the current paradigm these "fixes" cannot keep up. This points to the need to invent new strategies to deal with today's unprecedented pace of new ICT introduction.

The need for this issue to be addressed is becoming better recognized both within the community of persons with disabilities and in the worldwide stakeholders community. Perhaps the best recent example is contained in the Convention on the Rights of Persons with Disabilities [13]. This convention, among many other important issues, specifically calls for the accessibility of information and communication technologies and related services (see, e.g., Articles 9 and 21). As of July 2009, 62 countries have ratified this convention. Currently, Canada recently ratified and the US is in the process of working towards ratifying this convention [12].

In this chapter, a few examples that demonstrate the accessibility barriers that are caused by mainstream ICT are given and three possible strategic areas that could work to solve this emerging problem are outlined. One of these strategies is the development of regulations and standards. This area will be discussed in greater detail with an attempt to make the case that appropriate technologies for people with disabilities will only be reliably available, at least in the relative near term (next 5–10 years), with regulation and the enforcement of standards that effectively mandate their deployment.

10.2 Examples of ICT: A Barrier or an Enabler

As discussed above, as emerging ICTs become generally available in our society, the trend also opens up the potential for persons with disabilities to take part in a wider range of activities. A few years ago, the Neil Squire Society [1] held a number of focus groups with persons with mobility and vision impairments to gain their input on public ICTs and ensuing challenges, barriers and potential wireless solutions. Among the outcomes was the general consensus that wireless technologies had promising potential to enhance inclusion. The top three services that were identified during this process were banking, retail (electronic payment) and transit.

There were many examples related to transit and banking including issues related to accessing automated transit ticket systems or banking machines. The proliferation of public services using automated dispensing machines (kiosks for banking, buying transit tickets, getting information, etc.) is revealing common barriers to persons with disabilities; however, if the machines are improved, the services can be transformed into opportunities for enhanced accessibility. For a detailed discussion on the use and accessibility of these technologies, see Ripat et al. [10].

Given the interest in electronic payment (often referred to as e-commerce) by persons with disabilities as an enabler if it were accessible, and its growing use and deployment [6], the Neil Squire Society recently carried out a large study on the accessibility of mobile payment systems with the blind, the hard-of-hearing, the deaf, and persons with mobility impairment [5]. The study highlights 12 key accessibility issues that affect the four different transaction methods that were studied (for more details on how these were selected, see Lew et al. [5]). A few examples of the barriers encountered were: voice menu systems that did not allow sufficient time for users with mobility impairments; voice menu systems that did not have a text-based alternative to allow access for deaf users; and web-based systems that did not adhere to all the Web Content Accessibility Guidelines [7], which made them largely unusable by the blind. For a video overview of some examples, including those noted above, of persons with severe mobility impairments using existing and potential emerging ICTs and related services see the website in [8].

10.3 Three Strategies

There are at least three main strategic approaches to improve accessibility of emerging ICTs: (1) working with industry, (2) developing government regulations and related standards, and (3) educating students who will design future ICTs.

1. Work with industry The not-for-profit sector that has specific domain expertise with persons with disabilities and technology needs to engage with industry to help them better understand the design challenges and potential solutions. The key is to help companies better understand their role, whether it be from a business case point of view or from the requirement to meet regulations, or both. Despite the fact that around the world there are a number of not-for-profit organizations that are attempting to engage with industry, the relationships have generally been very difficult, with only a few examples of what most would consider minor successes. Neufeldt et al. explores these challenges in greater detail [9]. Some of the ways that the not-for-profit sector can engage with industry are: consumer focus group facilitation; needs determination studies; usability and accessibility evaluations; sophisticated simulation studies; device development and adaptation; accessible design consultation; usability guideline and standards development; and demonstrations of proof-of-concept solutions in real and/or simulated environments. Given that most not-for-profit organizations do not have the capacity to do this for free, industry needs to understand the benefit for them to support these kinds of engagements and therefore retain these not-for-profits on a fee-for-service basis. Currently, few industries have been ready to seriously entertain a business case to do this work, although with the growing aging demographics there are signs that some industries are starting to consider this in their business planning. To date, most industries that have considered some form of inclusive design have been more motivated to do so by the existence or threat of regulation requiring them to do so.

2. Engage proactively in the development of government regulations and related standards The work by Stienstra et al. [11] is but one example of research that has pointed out the importance of establishing government enforceable regulations and their related standards. See also G3ict, the Global Initiative for Inclusive Information and Communication Technologies [4], which is a flagship advocacy initiative of the United Nations Global Alliance for ICT and Development. Initiated in December 2006 by the Wireless Internet Institute, G3ict is a public-private partnership dedicated to facilitating the implementation around the world of the Digital Accessibility Agenda defined by the Convention on the Rights of Persons with Disabilities. This initiative strongly supports the implementation of appropriate government policy with heavy emphasis on the coordinated development of standards regarding accessibility of ICT. This initiative involves government, industry and nongovernment organizations. Stienstra et al. [11] also provide examples of various pieces of legislation from around the world (primarily the US, Australia, and the UK) that in some way regulate accessibility of ICT. Some of the most commonly cited relevant pieces of legislation in the US are: Section 501 of the Rehabilitation Act [18] and Section 255 of the Federal Communications Commission (FCC) Communications Act [17] (both currently under review) [15].

3. Education of students Ensuring that our postsecondary students, primarily engineers, computer scientists, and industrial designers, who will be involved in the design of new technologies, are taught the principles of inclusive design is a long-term approach, but one that may pay off as a sustainable solution to the production of accessible ICT in the consumer mass market. Ideally these students should not just be taught the theory but also have the opportunity to complete projects that directly involve persons with disabilities and their access challenges with ICT. The students will then carry these experiences into their workplaces with the grounded realization that inclusive design results in a much broader range of people being able to use the technology, thereby increasing a company's market share for a given product while also helping to contribute to equal access for all.

10.4 The Need for Regulations and Standards

Due to the lack of voluntary uptake by industry as noted above, regulations and standards will play the key role in providing substantial accessibility of emerging ICTs in the near-term, over the next 5–10 years. For instance, officials in the EU acknowledge that there is a clear correlation between the existence of legislation and the actual level of progress on accessible ICT [3]. The EU has been exploring a general legislative approach, but there is not yet a clear consensus on possible specific EU legislation dedicated to accessible ICT, including elements such as scope, standards, compliance mechanisms, and links to existing legislation [3].

In Canada, as another example, there are no regulations related to the accessibility of ICT. However, the Canadian Radio-television and Telecommunications Commission (CRTC) recently held hearings on accessibility related to both communication and broadcast technologies in Canada. These hearings covered a range of issues, and the CRTC made several rulings and recommendations as a result [2]. An example of one of the directives that relates to wireless handheld technologies by Wireless Service Providers (WSPs) is [2]:

> Accordingly, the Commission requests that, by 21 October 2009, all WSPs offer and maintain in their inventories at least one type of wireless mobile handset that will provide access to wireless service by persons who are blind and/or have moderate-to-severe mobility or cognitive disabilities.

It is illustrative that regulators are starting to pressure industry more to produce accessible ICT. While time will tell, this ruling (which is currently a "request"), could be one of the more progressive pieces of regulation to make industry take into account inclusive design. If industry does not respond to this "request", then the real test will be whether the CRTC will create an updated directive that requires companies to offer accessible technologies. This type of requirement (with sanctions) will produce more procurement demand (and in this case, new demand in areas such as access for persons with severe mobility impairment that have not yet seen specific regulation) by the WSPs on the manufacturers of handsets. The handset manufacturers will be forced to start producing at least some of their products in a format that is accessible and/or is easily made accessible for the targeted consumers.

The US currently has what is perhaps one of the best examples of regulation driving the deployment of mass market inclusive technology: the FCC's change to the Hearing Aid Compatibility Act of 1998 (HAC Act) [16]. The change, introduced in 2003, required wireless handset manufacturers and WSPs to make a substantial part of their inventory compatible with hearing aids. Through a ramping-up process starting in 2008 and finishing in 2011, the change has resulted in the almost universal availability of cellular telephones that are hearing aid compatible in the US, and because of the large ripple effect of the US market, these wireless telephones are also available in many other jurisdictions as well. Now a person with a hearing disability who uses a hearing aid can simply buy the phone, on a level playing field with a hearing person, and have a piece of technology that works right out of the box. This is the ideal case, that regulation can bring about in the relative near-term: universally accessible or easily adaptable ICT for persons with disabilities.

As noted in the Introduction, the UN Convention may play a big role in getting jurisdictions in many parts of the world to establish appropriate regulations. In particular, the Optional Protocol to the UN convention will increase pressure on countries that ratify the Optional Protocol to establish mechanisms to ensure the accessibility of ICT. As quoted from the Convention [14]:

> The Committee on the Rights of Persons with Disabilities is a body of independent experts tasked with reviewing States' implementation of the Convention. The Committee periodically examines reports, prepared by States, on the steps they have taken to implement the Convention. For those States that are party to the Optional Protocol, the Committee also has authority to receive complaints from individuals of alleged breaches of their rights and to undertake inquiries in the event of grave or systematic violations of the Convention.

Therefore, in countries that sign on to the Optional Protocol, there will exist a robust regulatory mechanism that will allow individuals to lodge complaints to this international body. Hopefully, just the fact that countries know that they will potentially be held to account will encourage the establishment of effective local standards and regulations in a timely fashion. As more and more countries establish these types of regulations, there will be a resultant demand on the manufacturers of ICT to bring us closer to the ideal, as illustrated above in the example of the hearing aid compatible cell phone handsets, in which persons with disabilities can easily access mainstream technology that is designed to be accessible or easily adaptable for their personal access needs.

10.5 Concluding Remarks

The potential for greatly enhanced inclusion for persons with disabilities using emerging ICTs is clear, but unfortunately the new barriers they present are also clear. Students better educated in accessible design of ICT and not-for-profit companies that represent persons with disabilities working with the ICT industry are long-term strategies that can help change these new barriers into enablers. More importantly, in the shorter term over the next 5–10 years, the implementation of more regulations and standards governing the accessibility of ICT is required. Governments and the general public that produce the empowering political will must act now, or more persons with disabilities will become further marginalized in this ever-increasing and faster-evolving information age.

References

[1] Birch G (2006) Findings from research conducted by the disability and information technologies (Dis-IT) research alliance – Retail and public services. In: Hard-Wiring Inclusion: A Conference about Building an Accessible ICT World, Winnipeg, Manitoba, pp 73–82. http://www.dis-it.ca/2006si/2006-10-26.php#research

[2] Canadian Radio-television and Telecommunications Commission (2009) Broadcasting and telecom regulatory policy CRTC 2009-430: Accessibility of telecommunications and broadcasting services. http://www.crtc.gc.ca/eng/archive/2009/2009-430.htm

[3] Commission of the European Communities (2008) Towards an accessible information society. Communication to the European Parliament, the Council, the European Economic and Social Committee and the Committee of the Regions. http://eur-lex.europa.eu/LexUriServ/LexUriServ.do?uri=COM:2008:0804:FIN:EN:PDF

[4] G3ict (2010) Global initiative for inclusive information and communication technologies. http://www.g3ict.com

[5] Lew H, Leland D, Birch G (2009) Evaluation of mobile payment systems for people with disabilities. Research report, Neil Squire Society. http://www.neilsquire.ca/

[6] Lin KJ (2008) E-Commerce Technology: Back to a Prominent Future. IEEE Internet Computing 12(1):60–65, doi:10.1109/MIC.2008.10

[7] Mobile Web Initiative (2010) The web on the move. http://www.w3.org/Mobile/

[8] Neil Squire Society (2009) Wireless technologies for people with disabilities. http://www.youtube.com/watch?v=JrJA7glzIBA&feature=related

[9] Neufeldt A, Watzke J, Birch G (2007) Engaging the business/industrial sector in accessibility research: Lessons in bridge building. The Information Society, An International Journal 23(3):169–181

[10] Ripat J, Watzke J, Birch G (2005) Public information and communication technologies: Improving access or crating new barrier? OT Now pp 21–24, CAOT Publications Ace

[11] Stienstra D, Watzke J, Birch G (2007) A three-way dance: The global public good and accessibility in information technologies. The Information Society, An International Journal 23(3):149–158

[12] United Nations (2008) Ratification of the convention on the rights of persons with disabilities. http://treaties.un.org/Pages/src-TREATY-id-IV~15-chapter-4-lang-en-PageView.aspx

[13] United Nations (2010) Convention on the rights of persons with disabilities. http://www.un.org/disabilities/convention/conventionfull.shtml

[14] United Nations: Secretariat for the Convention on the Rights of Persons with Disabilities (2010) Frequently asked questions regarding the
convention on the rights of persons with disabilities. `http://www0.un.`
`org/disabilities/default.asp?id=151#sqc8`

[15] US Access Board (2008) Update of the 508 standards and the telecommunications act guidelines. `http://www.access-board.gov/sec508/`
`update-index.htm`

[16] US Federal Communications Commission, Consumer and Governmental
Affairs Bureau (2008) Hearing aid compatibility for wireless telephones.
`http://www.fcc.gov/cgb/consumerfacts/hac_wireless.html`

[17] US Federal Communications Commission, Consumer and Governmental Affairs Bureau (2008) Section 255: Telecommunications access
for people with disabilities. `http://www.fcc.gov/cgb/consumerfacts/`
`section255.html`

[18] US General Services Administration (1998) Section 508. `http://www.`
`section508.gov/index.cfm?FuseAction=Content&ID=17`

Chapter 11
Small Markets in Assistive Technology: Obstacles and Opportunities

Jaimie F. Borisoff

Abstract While the inherently small market for assistive technology (AT) can be a significant hurdle in the development of AT solutions, small markets can also provide opportunities to foster technology innovation under the right circumstances. Focusing on the wheelchair industry, I summarize the numerous obstacles that small companies face in trying to address small AT markets, as well as the range of opportunities available to assist small companies in their efforts to impact the quality of life of those with disabilities. Indeed, small market AT may be a perfect fit for the growing field of "small batch" manufacturing in combination with motivated individuals suddenly empowered by a host of new technologies. I close with a brief discussion of future possibilities for AT in small markets.

Assistive technology (AT) has seen great progress over the past half century, creating significant positive impact to overall quality of life for many people. However, that most obvious form of AT, the wheelchair, provides some muted historical perspective [4]:

> The wheelchair has, for most of its history, been a design that segregated instead of integrated.

An almost complete lack of innovation over several decades contributed to this situation. One reason for the disappointing performance by the wheelchair industry was the monopoly enjoyed by Everest and Jennings Inc. (E & J), which had up to 90% of the manual wheelchair market share until 1978 [9]. Perhaps the simple lack of competition never created the forces necessary for innovation in wheelchair technology. A US Government antitrust suit was

Jaimie F. Borisoff
Instinct Mobility Inc., and Brain Interface Lab, Neil Squire Society, International Collaboration on Repair Discoveries, 818 West 10th Avenue, Vancouver, BC, V5Z 1M9, Canada, e-mail: jaimieb@gmail.com

M.M.K. Oishi et al. (eds.), *Design and Use of Assistive Technology: Social, Technical, Ethical, and Economic Challenges*, DOI 10.1007/978-1-4419-7031-2_11, © Springer Science+Business Media, LLC 2010

finally successful in limiting the market dominance of E & J: by 1983, the market had expanded to comprise over 50 wheelchair manufacturers [9]. Twenty years after the antitrust suit, the number of wheelchair manufacturers had ballooned to over 170 companies [6]. The early wheelchair industry seemed to be simply one very large market served by the ubiquitous steel folding wheelchair manufactured by E & J. In reality, once the barrier to industry participation was reduced, the wheelchair industry matured into a collection of many small markets, often completely independent from each other. Small companies that were effective at creating innovative products also created their own opportunities in niche markets within the overall marketplace.

Today, several areas in AT, including the wheelchair industry, have mainstream companies with billion dollar revenues, large scale manufacturing resources, and international sales and distribution channels. However, much of AT serves customers with an extremely wide variety of disabilities, often encompassing unique needs that necessitate specific innovations and solutions that are only deliverable with small market economics.

This chapter first summarizes the obstacles that individuals, academia, and small companies face in trying to address small AT markets. Then, I present a range of opportunities available in small markets. Lastly, I discuss the brewing revolution in "small batch" manufacturing, which has the near-term potential to provide specific AT solutions in small markets. This seemingly perfect fit can address the specific needs and interests of individuals suddenly empowered with a host of new tools and services.

11.1 Small Market Obstacles

Small markets in any industry (including AT) present many obstacles to sustainable business growth. Unfortunately, there are also several obstacles that are more specific to AT and the medical device industry – these will be summarized following a brief discussion of the most widely applicable issues.

11.1.1 Obstacles Beyond AT

In some manner, almost all obstacles related to small markets involve money and resources. Bringing new products to market is simply an expensive proposition, both for start-up companies as well as for established companies adding to existing product lines. Introducing new products into a small market further emphasizes these cost issues.

A start-up company will find it difficult to raise money for its ideas that address a small market. A rule of thumb for pursuing venture capital is whether it is possible in the short-term (e.g., less than 5 years) to attain $100 million in annual revenue. Many successful AT companies that have been operating for decades have sales only in the low tens of millions of dollars. It is not uncommon for AT companies to have sales that are even smaller. Of the approximately 170 wheelchair manufacturers operating in the United States as of 2003, only five companies had sales in excess of $100 million, although the industry total reported revenue was $1.3 billion [6]. Thus the majority of established AT companies could not attract venture capital. Furthermore, starting up without adequate resources is one of the most detrimental factors in the long term success of new companies [3].

Similarly, an established company operating in small markets is faced with limited revenues and little free operating cash. Cash flow is the key resource for pursuing new product innovation – an endeavor often seemingly at odds with maintaining a profitable small business. As well, all companies operating in small markets must prioritize their employees, who are a prized resource. Often a lack of cash results in a lack of people with the expertise and time to mount a thorough and viable product development stream.

Some areas of product development and innovation that need ample cash and resources include:

- People and software for product design.
- Prototype development and design iteration.
- Intellectual property analysis and protection.
- Market testing and feedback.
- Manufacturing considerations.

Small market operations shape all of these issues. Are the people with relevant experience available for design? Does the company possess prototyping capabilities in-house or does the company need to source more expensive solutions? Do the projected revenues support patenting costs? If so, which geographical markets are realistic for a small company to successfully deliver the product to the end users? Does the company need to hire outside firms for market testing, and if so, how difficult will it be to gather enough customers for proper evaluation? Academic researchers face a similar problem when addressing specific medical problems (essentially a small market). For example, researchers in spinal cord injury have difficulty running clinical trials without expensive multi-centre collaborations across North America.

Manufacturing issues also comprise many of the obstacles in small market operations due to the inherently low volumes involved. Manufacturing a new product requires new tooling, processes, and training for the supplier. These costs are somewhat (though not entirely) fixed regardless of volume size, thus putting a greater burden on companies with only low volume sales projections. In addition, excess inventory in stocked parts and finished product for higher volume manufacturing incurs a non-negligible expense. The alternative to this expense is for the company to absorb a higher per-unit item cost by running an operation with minimal inventory. However, this strategy may incur further costs by increasing customer delivery times and also has the disadvantages associated with back-ordered products.

Sales and distribution is a great challenge for companies in small markets. Mainstream products in large markets typically use large-scale distribution channels (e.g., nation-wide drugstore chains for simple crutches and canes). A company selling a product in a large market can usually target sales channels, each with many outlets, all handled by a single representative. On the other hand, a company selling a product in a small market must establish relationships with distinct retail outlets in each geographical region – this is a considerably more expensive and time-intensive process than is required in larger markets.

11.1.2 AT-Specific Obstacles

Issues related more specifically to the AT industry include (1) additional people in the supply chain, (2) education and training, (3) regulatory issues, and (4) sales support and ongoing servicing and maintenance.

The supply chain of many businesses (e.g., bicycles) includes the manufacturer, distributor, retailer, and customer. The AT industry (e.g., wheelchairs) also includes the clinician or prescriber as well as the funder. The process of purchasing an AT device often takes several months and includes many people besides the customer: a clinician and/or therapist, a funding agency representative, a dealer salesperson, and possibly a manufacturer's representative. A company introducing a new product into this group must educate and train many people to promote product sales effectively. The more innovative and different that product is, the more difficult the educational effort will be. Furthermore, industry professionals often act conservatively when faced with new products, preferring to recommend established companies and products until the newer products have been proven over time. This pattern exacerbates two difficulties for small market companies: long periods of time with low revenues and the burden of sales education. Low revenues make it difficult to support the sales and education efforts. The burden of education is even greater when an innovative product must break into a market comprised of united force (e.g., many competitors all selling the same type of competing product) – a difficult situation for any company. Even if the company successfully completes the hard work of proving an innovative product and establishing its market niche, it is entirely possible that larger competitors may then enter the maturing niche market with significantly reduced risk. The payoff of the innovating company's efforts could be claimed by its mainstream competitors.

Most AT devices are developed and funded based on medical necessity. Since medical devices typically fall under the jurisdiction of the FDA or its equivalent, most AT devices are required to pass a regulatory hurdle, another expensive and often time-intensive requirement. In the case of wheelchairs,

FDA approval typically requires certification to RESNA standards [5]. Some jurisdictions make this more cumbersome by using their own rules and processes to certify a new device. A small market AT company needs to prioritize these jurisdictions to focus on the ones in which they have established strength (e.g., through sales channels and relationships with clinicians). Navigating the regulatory process requires money, people, and product samples, all of which can be scarce commodities in small market companies.

Devices considered AT are usually more complex than mainstream products. Consider, for example, a powered wheelchair with custom seating as opposed to a mountain bike. AT products require ongoing support and servicing that often only specialists can perform. Funding agencies often insist on long term support contracts with retailers, an additional burden for small manufacturers with unique products.

Finally, demographics may play a transformative role in the AT industry in the coming years. Similar to many small businesses, the founders of many AT companies are nearing retirement. If in-house heirs are not forthcoming, will the pool of outside investors commit to make significant acquisitions and continue supplying much needed products to these niche markets? The nature of these poorly understood businesses, often with marginal growth prospects, may force many products out of the market. Hopefully, these obstacles will create opportunities elsewhere for other small innovative companies to prosper.

11.2 Small Market Opportunities

The above obstacles notwithstanding, small markets in the AT industry also offer opportunities to small, nimble, and innovative enterprises.

Necessity often drives the development of new innovations, and there are certainly examples of this in AT. Even the founder of the early monopolistic Everest and Jennings wheelchair company used a wheelchair. Highly motivated people who clearly understand the problem will hopefully develop better products. User-driven design is often paramount in small market innovations, and seen not only in companies (e.g., Marilyn Hamilton of Quickie Wheelchairs), but also in academia and in non-profit organizations. For instance, The Tetra Society and CanAssist are staffed by engineers and technicians able to create solutions to many problems. However, the effort is only undertaken once an end-user has identified a problem and initiated a solution. Sophisticated customers thus contribute to the innovation process. Resulting solutions are then made available to others when possible and contribute to the overall technology pool of available AT solutions. There are rarely issues surrounding intellectual property or proprietary technology that prevent dissemination of new AT solutions developed in this manner. Big businesses are simply unable to easily contribute to the overall community innovation effort in this way.

Similarly, other people involved in the supply chain are often motivated to better understand the nuances of niche markets. It is possible to find amazing sales people and therapists with years of experience and understanding within a particular small market. Such people are a valuable source of evaluation and feedback for small market innovators and usually come at little or no charge. A therapist, for example, benefits from the availability of better solutions for the therapist's clients.

From an economic standpoint, small markets often offer less competition. Thus, a slow, inexpensive route to commercialization does not carry the same predatory risks from big companies since the eventual payoff is not enticing to them. It is also possible to innovate and commercialize a new product without patent protection in a marketplace with less competition. Innovators can fly "under the radar" much more easily in small markets.

A successful new product in larger markets may eventually become the victim of its own success. As a new market matures and becomes mainstream, the innovative company may be overwhelmed by larger competitors. However, the nature of most small AT markets ensures that this will not happen – usually, mainstream growth is simply not possible due to the small population size of the target consumers.

Large companies can be at a disadvantage due to their size, especially when developing and introducing new products to big markets. A large company has an extensive supply chain and ongoing commitments for its own existing products. Huge inventories of parts and investment in manufacturing can lead to products that are long overdue for a facelift. Thus, a large company in a large market may not be able to be responsive to newly identified problems or product features coveted by customers.

A small market company has two advantages in this case. First, the company can introduce products without potentially cannibalizing its own existing product lines. Second, low volume manufacturing enables quick redesign of specific parts, or even termination of a product in order to introduce a completely new one.

Larger companies may hesitate to innovate because of the possibility of confusing customers and partners. This is especially the case in AT, with numerous people involved with purchasing a product. New, complementary products, or products with new features too similar to existing products, create too many messages about the best solution for a particular problem. Creating more products for the same market may lead to lower volumes for each product – rendering a large company without its normal advantages over those used to operating in small volume markets.

Other revenue and cost advantages that are sometimes present in small AT markets are found in product pricing and marketing efforts. Innovative products in small AT markets may command higher margins, hence the payoff for low volumes is a higher per-unit profit. Regarding cost, it may be easier to introduce a small-market AT product because of the ease of publicizing in the restricted avenues for a particular niche. Therapists and customers have familiar sources of information for new products in their particular niche. The Internet makes publicizing new products even easier, with online communities dedicated to particular disabilities.

Fundraising groups dedicated to specific disabilities are often quite focused and effective at raising money for research, or at providing specific AT solutions. For example, the Christopher and Dana Reeve Foundation, the Rick Hansen Institute, and the National Federation of the Blind have lobbied governments and raised funds for research and technology transfer. Companies looking to innovate in these areas may benefit from such organizations, with or without traditional partnerships with academia. Targeted grants are sometimes available for specific research into small market AT solutions, including joint industry-university funding. There is also regulatory assistance available in certain circumstances, for example, through the FDA's Humanitarian Device Exemption. Thus, non-traditional resources are often available to help innovators make an impact in small AT markets. Hopefully, these mechanisms create enough incentive to further technology innovation.

11.3 New Opportunities in Small Market Innovation

The term assistive technology may be redundant, perhaps better described as simply technology for individuals with a disability (see Ladner [8, 7] for more on this issue of redundancy and individual empowerment). Rick Hanson, Terry Fox, and other individuals have helped raise international awareness of the capabilities of persons with disabilities. We are moving to a "social model" of disability and away from a "medical" or "rehabilitation" model. That is, people with disabilities (and an aging population) are part of the diversity of life, not necessarily in need of treatment and cure. But people with disabilities do need full access to all facets of community, with complete dignity and integrity when at all possible. User-driven AT design for small markets is closely related to issues of individual empowerment within this social model. It is possible that technology innovation may greatly support the efforts of individuals to create their own novel solutions for their own specific wants and needs, thus further encouraging small market innovation efforts.

As described in Sect. 11.2, user-driven design is a powerful innovation force. Computers and the Internet have now opened up countless new opportunities for individual users of almost anything. This is most obvious in terms of software and information. As Chris Anderson of Wired Magazine [1, 2] explains, new opportunities abound for small scale manufacturing and product development, a perfect fit for AT innovations. This is an application of the "long tail" of stuff [2], with do-it-yourself manufacturing, crowdsourcing, and micro-factories. A small market innovator with very little money can use open-source design tools to design a novel device. Factories around the world, and particularly in China, now accept small batch and prototype orders from anyone via the Internet. In a matter of days, the innovator's AT solution arrives via post for evaluation and design iteration. Post your design to your community and let them help create the ideal solution. If the solution works for one individual, chances are that it will work for others.

What does the future hold for technology innovations for persons with a disability? Who knows? But a safe bet would be that an individual with a disability is behind the innovation effort.

11.4 Afterword

The author of this chapter has been involved with spinal cord injury and assistive technology research for over 10 years and is a wheelchair user. He recently started his own small business to bring novel wheelchair technology to the marketplace. This real-world experience, in combination with discussions with several colleagues who have backgrounds in the AT industry (see Acknowledgments), shaped much of the commentary in this chapter.

Acknowledgements The following individuals have each contributed AT innovations to the marketplace: Murray Slagerman of Ki Mobility LLC, Dr. Arthur Prochazka of the University of Alberta, Dr. Mark Richter of Max Mobility LLC, and Harry Lew of the Neil Squire Society. I would like to thank them for their very helpful comments on this chapter.

References

[1] Anderson C (2008) The long tail. Hyperion, New York
[2] Anderson C (2010) In the next industrial revolution, atoms are the new bits. Wired Magazine. http://www.wired.com/magazine/2010/01/ff_newrevolution/
[3] Baldwin J, Gray T, Johnson J, Proctor J, Rafiquzzaman M, Sabourin D (1997) Failing concerns: Business bankruptcy in Canada. Tech. Rep. 61-525-XPE, Statistics Canada, Micro-Economic Analysis Division, Ottawa, ON
[4] Cooper R (1998) Wheelchair selection and configuration. Demos Medical Publishing, New York
[5] Cooper R (2006) Wheelchair standards: It's all about quality assurance and evidence-based practice. Journal of Spinal Cord Medicine 29(2):93–94
[6] Cooper RA, Cooper R, Boninger M (2008) Trends and issues in wheelchair technologies. Assistive Technology 20(2):61–72
[7] Ladner R (2010) Personal communication

[8] Ladner R (2010) Accessible technology and models of disability. In: Oishi M, Mitchell IM, Van der Loos H (eds) Design and Use of Assistive Technology, pp 23–29. Springer, Berlin

[9] Shepard DS, Karen SL (1984) The market for wheelchairs: Innovations and federal policy. Health Technology Case Study 30, Congress of the United States, Office of Technology Assessment, Washington, DC

Index

9 781489 989802